Modern Researches in Metallurgical Engineering

Modern Researches in Metallurgical Engineering

Edited by **Darren Wang**

New York

Published by NY Research Press,
23 West, 55th Street, Suite 816,
New York, NY 10019, USA
www.nyresearchpress.com

Modern Researches in Metallurgical Engineering
Edited by Darren Wang

International Standard Book Number: 978-1-63238-334-1 (Hardback)

Printed in the United States of America.

Contents

Preface

Metallurgical engineering is the science of innovation, implementation and modeling of the process that transforms metals, alloys and other engineering materials via cost-effective methods. Since the Bronze Age, i.e. 3000 years ago, metallurgical engineering has played a pivotal role, wherein alloys and metals were shaped to build weapons and tools. This science has evolved over the ages. This book covers an array of latest theories and practices in the field of metallurgy and all related areas, including mineral processing, extraction, thermal treatment and casting.

This book is a result of research of several months to collate the most relevant data in the field.

When I was approached with the idea of this book and the proposal to edit it, I was overwhelmed. It gave me an opportunity to reach out to all those who share a common interest with me in this field. I had 3 main parameters for editing this text:

1. Accuracy – The data and information provided in this book should be up-to-date and valuable to the readers.
2. Structure – The data must be presented in a structured format for easy understanding and better grasping of the readers.
3. Universal Approach – This book not only targets students but also experts and innovators in the field, thus my aim was to present topics which are of use to all.

Thus, it took me a couple of months to finish the editing of this book.

I would like to make a special mention of my publisher who considered me worthy of this opportunity and also supported me throughout the editing process. I would also like to thank the editing team at the back-end who extended their help whenever required.

Editor

Part 1

Metal Recovery from Industrial Waste

Possibilities of Exploitation of Bauxite Residue from Alumina Production

Marián Schwarz[1,2] and Vladimír Lalík[1]
[1]*Technical University in Zvolen, Faculty of Ecology and Environmental Sciences,*
[2]*Institute of Chemistry, Center of Excellence for White-green Biotechnology,*
Slovak Academy of Sciences, Nitra,
Slovak Republic

1. Introduction

The world aluminium production was 31.9 million tons in 2005 (Hetherington et al., 2007) and the demand is still growing. Aluminium is obtained from aluminium ore called bauxite in compliance with the locality where it was found near Les Baux in southern France in 1821. Bauxite can be found on all the continents, except for Antarctica, while the most abundant deposits are situated in tropical and subtropical zones, which provide optimal climatic conditions to its formation (Bardossy, 1990).

From ecological point of view the most significant world problem in aluminium production is creation of considerable amount of insoluble bauxite residue (BR). It is waste that, due to way of its production, is determined as red mud (RM) or brown mud (BM). 1 t of produced aluminium gets 2 t of waste and it is estimated that nowadays, during the bauxite processing, 120 million t of BR is produced worldwide (Red Mud Project, 2011).

Even if there is all-out effort in the world to reuse the waste from the aluminium production (Paramguru, 2005; Snars & Gilkes 2009; Liu et al., 2011; Maddocks et al., 2004), its majority part is dumped and due to its dangerous properties (high pH value, strong alkalinity, increased content of radioactive substances etc.) it is a significant environmental burden now and it will be a load also in the future. All world aluminium producers try to solve above mentioned problem more or less successfully. In this chapter, we propose a brief overview of the most significant possibilities of waste mud exploitation together with assessment of possible influence on the environment based on ecotoxicity tests.

2. Waste from alumina production

Many aluminium works are not located near to bauxite deposits. Consequently, they have to import bauxite or they buy pure aluminium oxide and the waste dumps stay at the localities of deposit or factories producing Al_2O_3. Chemical composition of bauxite differs significantly in dependence on the locality of its deposit. Aluminium, as its most important component, occurs in bauxite in the form of hydrated aluminium oxide, whose content varies, while only ores with Al_2O_3 content more than 65 % are economically interesting. In dependence on chemical composition, physical properties of bauxite also vary, e. g. colour (from yellow-white to grey, from pink to dark-red or brown) or structure (earthy, clay or compact ore).

Bauxite refining for Al_2O_3 production can be realized via several procedures (alkaline, acid, thermic, or high-pressured). However, the most spread procedure is the one of Austrian chemist K. J. Bayer, first patented in 1888 (Sintering process) and second in 1892 (Bayer process). Comparison of both processes is shown in Figure 1. (Klauber et al., 2009). The Bayer process resides in bauxite leaching by NaOH at increased pressure and temperature, in clarification of dissolved sodium aluminate and consequently in precipitation and calcination (Hind et al., 1999). Aluminium oxide, produced by the Bayer process, is relatively clean. It contains just several hundredths of per cent of impurities (oxides of iron and silicon). Waste is called red mud due to its intensive red colour. It is a suspension of very fine solid particles (more than 90 vol. % is lesser than 75 μm) and solid concentration ~400g.l^{-1}.

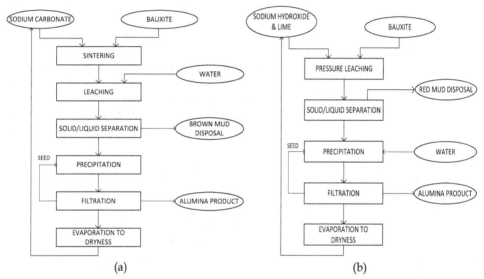

Fig. 1. Alumina production flowsheets of Sintering (a) and Bayer process (b) (Klauber at al., 2009).

Bauxite of higher silicon content is not suitable for the Bayer process, so in this case aluminium oxide has to be produced by so-called sintering process. For instance, in aluminium work in Žiar nad Hronom in Slovakia, 4.548 million t of Al_2O_3 were produced and 8.5 million t of waste mud emerged from 1957 to 1997[1] (Fig. 2.). Waste from the sintering process is called brown mud. In comparison with red mud it is more solid, because it contains higher portion of solid component, but it has lower content of Fe_2O_3 and Na_2O.

Similarly to the change of chemical and mineralogical composition of bauxite in dependence on its deposit, the composition of waste mud changes, too. Predominate component of waste mud are oxides of iron in the form of crystallic hematite (Fe_2O_3) that is the source of red colour of bauxite, or goethite FeO(OH). Aluminium oxide in the form of boehmite (γ-AlOOH) has its important place, and oxides of Ti, Na, Si, Ca, K and other metals,

---

[1] In this year was alumina production terminated in Slovakia and ready alumina began to import.

Fig. 2. Bauxite residue (brown and red mud) dump in aluminium work in Slovakia before (left, 2006) and after (right, 2011) revitalization.

predominantly in the form of quartz, sodalite, gypsum, calcite, gibbsite, rutile, represent other minority components. In Tab. 1. chemical composition of waste mud from several world aluminium oxide producers is shown. All the data, except for Slovakia, is related to waste mud originated by the Bayer process (red mud).

Origin of Al_2O_3 producent	Fe_2O_3	Al_2O_3	TiO_2	Na_2O	SiO_2	CaO	CO_2	LoI*
France	19.4	27.3	1.9	5.2	10.2	8.6	-	16.5
Greece	41.9	14.8	5.8	10.0	10.0	-	-	-
Turkey	37.5	19.3	5.1	9.8	16.4	2,3	2.1	7,4
Guinea	56	21.1	2,3	0.4	3,7	0.2	0,8	15.7
Egypt	31	20	21	6	8	4	-	-
Russia	41,6	14.2	3,8	3.3	9,2	12.0	6,6	11.9
India	31.9	20.1	21.2	6.5	8.5	3.0	-	8.5
USA	41.6	16.4	5.5	6.8	1.3	6.7	-	-
Hungary	33-48	16-18	4-6	8-12	9-15	0.5-3.5	-	-
Brazil	43.7	16.8	4.0	9.8	-	-	-	-
Slovakia	45	15	6.0	7.5	13	2	-	6
Slovakia**	32	12.5	3.3	4.0	11.5	24	-	11
China	27.9	22.0	2.3	10.5	20.9	6.23	-	10.0

Note: *LoI – Lost of Incineration, **data related to brown mud

Table 1. Chemical composition of bauxite residue (BR) by origin of Al_2O_3 producer (in wt. %). Modified from article written by Kurdowski & Sorrentino (1997).

From ecological viewpoint it does not matter whether it is red or brown mud, because they both are dangerous waste and environmental load. The danger resides mostly in high alkali content, where pH values in fresh mud are higher than 13. There is a risk of percolating into groundwater, in case of rain storm there is a danger of dam crash (e. g. in Hungarian aluminium work Ajka in 2010), during the dry period of the year it is fly of dust particles and air pollution by aggressive aerosol and blocking large surface by dangerous waste.

Except for quoted predominant components, lesser amount of other metals, e. g. Na, K, Cr, V, Ni, Ba, Cu, Mn, Pb, Zn, Ga, Tl etc. can be found in waste mud.

Specific surface of waste mud reaches relatively wide values in the extent of l0 - 21 $m^2.g^{-1}$ and density of 2.51 - 2.7 $g.cm^{-3}$. 1,4 % from original waste mud is dissolved in water, but 12,3 % is dissolved in hydrochloric acid (Kurdowski & Sorrentino, 1997).

Apart from above mentioned inorganic compounds in waste mud, the presence of many organic substances was described in literature (Hind et al., 1999) – such as polyhydroxyacids, alcohols and phenols, humic and fulvic acids, saccharides, sodium salts of acetic acid and oxalic acid etc. The presence of above mentioned organic substances is related to decay products of vegetation that grew up on the sludge beds and concerning enormous amount of red mud it is insignificant.

Concerning anticipated use of waste mud from Al_2O_3 production, especially as a part of construction materials, intensive research of its radioactivity was carried out, too. In several works (Hind et al., 1999; Somlai et al., 2008; Idaho State University, 2011) there was found out that total radioactivity of bauxite and of waste mud as by-product of bauxite processing is several fold higher than background. Therefore it is necessary to examine radioactivity level in products, in which waste mud from Al_2O_3 production was used as additive, while its part by weight would not exceed 15 %. In Tab. 2 there are average, minimal and maximal values of radioactivity for ^{226}Ra, ^{232}Th and ^{40}K of Hungarian authors (Somlai et al., 2008), who present that activity concentrations of original minerals can considerably change in the samples of bauxite and red mud during their chemical treatment, while no significant differences were marked in samples taken from the surface or from inside of the material.

Sample	Values	Activity concentration (Bq.kg⁻¹)		
		^{226}Ra	^{232}Th	^{40}K
Bauxite	average	419	256	47
	min - max	132–791	118–472	10–82
Red mud	average	347	283	48
	min - max	225–568	219–392	4.9–101
Natural background (soil)*	average	48	45	420

Note: *data acquired from Idaho State University (2011)

Table 2. Radioactivity level of bauxite and red mud (Somlai et al., 2008).

As it follows from Tab. II in case of ^{232}Th and ^{226}Ra the measured values reach 5 – 7-fold of the background, while in case of ^{40}K they stay at lower value. Similar results were presented also by other authors undertaking radiological research (Akinci, A. & Artir, R. , 2008). Waste mud depositing is connected with high fees for the depositing, because the deposit site must be constantly monitored for the ecological threat of possible groundwater contamination.

3. Exploitation possibilities

It is evident from the studied literature (Kurdowski & Sorrentino, 1997; Andrejcak & Soucy, 2004; Liu, Y. et al., 2011; Paramguru et al., 2005; Bhatnagar et al., 2011; Klauber et al., 2009) that exploitation possibilities of red mud for next processing are considerable, though just a few of described techniques were put into practice, mainly because of economic reasons. These areas can be divided into several basic categories that include thermic, hydrometallurgical, mechanical, magnetic and chemical separation processes summarized in Tab. 3.

Areas of treatment	Processes	Product
metallurgical industry	drying, reduction	production of steel and other metals (Fe, Al, Tl, Ga, V etc.,) melting agent
construction industry	drying, sintering, annealing	production of Al_2O_3 and cements
	addition	production of concrete fillers
	drying, pressing, firing	bricks production
	drying, granulation	aggregates of special concretes
glass and ceramic industry	addition	glass manufacturing
	drying, pelletisation, annealing	ceramic manufacturing, ceramic glazes
chemical industry	chemical treatment	catalyst
	drying and chemical treatment	production of adsorbents, pigments, resin contents, filler for plastics
agriculture	addition to soil	improvement of soil physicochemical properties
	neutralization, adsorption	remediation of soil, revegetation
water supply	adsorption	removing of moisture and other undesirable components, treatment of liquid wastes
	coagulation	coagulant
other	drying and chemical treatment	filtration material
	adsorption	neutralization of acid mine drainage

Table 3. Overview of BR application areas from the alumina production.

3.1 Metallurgical industry

Beside material recovery of waste mud, the possibilities of sodium returning into the production process of aluminium oxide were studied on the base of above stated chemical and mineralogical composition (Lamerant, 2000). Processes for the recovery of other metals or of their oxides included in waste mud, e. g. Fe (Li, 2001; Piga et al., 1995; Qiu et al., 1996; Thakur & Sant, 1983), Al_2O_3 (Ercag & Apak, 1997), resp. TiO_2 (Agatzini-Leonardou et al., 2008), individually and in combination are equally important. Chinese authors describe nearly non-waste technology of red mud processing (Liu W. et al., 2009). They studied deep-drawing quality of iron by direct reduction in the sintering process and subsequent magnetic separation under different conditions (temperature and time of sintering, carbon –

red mud ratio and content of additives) and, in the next step, use of aluminosilicate residues in building industry for bricks production, where the problem of soluble sodium salts was solved by their calcification to insoluble aluminosilicates.

The possibility of use of minority components in waste mud still remains interesting. However, in spite of the existence of several patents the economics of the whole technology is decisive. The possibilities of recovery of rare-earth metals (Sc, U, Ga, Tl, V, Ce, Y etc.) were described (Smirnov & Molchanova, 1997; Ochsenkühn-Petropoulou et al., 2002; Gheorghita & Sirbu, 2009), which, in most cases, reside in cementation of waste mud, its extraction in acid medium and finally by the use of ion-exchange methods. Relatively high scandium content in bauxite residue processed by Aluminium of Greece allowed extraction of the Sc by nitric acid leaching followed by ion exchange separation (Ochsenkühn-Petropoulou et al., 1995; 1996). Obviously no impact was observed on the amount of residue to be stored by extraction of only minor component from bauxite residue.

Extraction of alumina and sodium oxide from red mud was envestigated (Zhong et al., 2009) by a mild hydro-chemical process and the optimum conditions of Al_2O_3 extraction were verified by experiments as leaching in 45% NaOH solution with CaO-to-red mud mass ratio of 0.25 and liquid-to-solid ratio of 0.9, under 0.8 MPa at 200 °C for 3.5 h. Subsequent process of extracting Na_2O from the residue of Al_2O_3 extraction was carried out in 7% NaOH solution with liquid-to-solid ratio of 3.8 under 0.9 MPa at 170 °C for 2 h. Overall, 87.8% of Al_2O_3 and 96.4% of Na_2O were extracted from red mud. The final residues with less than 1% Na_2O could be utilized as feedstock in construction materials.

Of the all amount of bauxite residue applications, only 15 % relate to metallurgical industry (Klauber et al., 2009) and of this amount about 30 % is related to steel making and slag additives. Bauxite residue is used as a source of aluminium, silicon and calcium to modify the properties of the slag to improve separation, setting and other qualities.

3.2 Construction industry

The most widespread use of red or brown mud processing is in construction industry (production of building components – bricks, concrete blocks, ceramic materials etc., concrete aggregate, special cement etc.). Brown mud obtained by sintering calcination method contains suitable reactive components, e. g. β-2 CaO.SiO_2, so it can be used (unlike red mud) for direct bricks production (Liu et al., 2009), where it is pressed into blocks and is calcined at high temperature, while compression strength of produced bricks reaches value of 1.9 MPa (Amritphale & Patel, 1987). In the second method red mud is mixed with binders and consequently it is hydraulically hardened or calcined at low temperature. Bricks from suitable mixture of clay, red mud and fly ash, calcined at 1000-1100 °C, reach high compression strength (40 to 70 MPa) (Dass & Malhotra, 1990) and they are suitable material for multi-storey buildings construction – due to their low moisture absorption, suitable density, exceptional fire resistance and characteristic colour and texture (Dimas et al., 2009).

Due to suitable Fe oxides-Al oxides ratio red mud is used in Portland cement production (Satish, 1997; Tsakiridis et al., 2004) or as additive in special cements production (Singh, et al., 1996). In both cases the addition of red mud into cement is limited especially due to strong alkalinity and it does not usually exceed 15 wt. %. According to the Japanese patent cement cinder was produced by mixing of calcium materials with red mud and subsequently by compressing and calcining in tunnel furnace at the temperature of 1300 - 1450 °C (Ogura, 1978). Natural red or red-brown colour of waste mud imitates the colour of brick and it is used for creation of interesting architectonic effects, for colouring concrete compounds, while colour adjustment of red mud can be done by pH adjusting, mixing with other oxides or by mild calcining (Satish, 1997). Replacing one-third of the content of silicate fractions in concrete compounds by red mud results in the fact that hardened concrete gains greater compression strength than concrete obtained by the use of quartz sand (Buraev & Kushnir, 1986). Hardening of concrete compound by utilization of red mud in amounts of 1 to 15 % under higher pressures (Kohno et al., 1998) helps to improve some final product properties, e. g. compression strength, water resistance, fire resistance etc. Several possibilities of the use of such concrete compounds were described, e. g. a filling for construction materials, materials used in dump construction, production of concrete blocks filling the dam structures or filling of building constructions (Klein, 1998; Kane, 1979; Di San Filippo, 1980; Browner, 1995).

3.3 Glass and ceramic industry

Red mud belongs to the group of so-called pozzolanic materials, which, after mixing with lime in the presence of water, harden and they form stable and durable compounds. The additions of red mud into clay compounds were described in the production of ceramic glass (Wagh & Douse, 1991; Sglavo et al, 2000a; Yalcin & Sevinc, 2000; Singh et al., 1996; Pratt & Christoverson, 1982). The content of red mud did not affect the sample porosity, but more deflocculated system originated, in which critical moisture content was increased. By augmenting red mud content to 20 %, after calcination at the temperatures of 950 and 1050 °C, growth of density and flexural strength were noticed in final ceramic products. It was brought about by bigger amount of glass share at bigger red mud content, which moreover by its natural colour replaces pigments addition to obtaining red-brown tones (Sglavo et al, 2000b).

3.4 Chemical industry

In organic synthesis red mud can be used as hydrogenation catalyst after its activation e.g. in hydrogenation of naphthalene to tetraline (Pratt & Christoverson, 1981). After several modifications (e.g. sulphidization) it is used as catalyst in petrochemical industry (Eamsiri, 1992; Álvarez et al., 1999), in dehydrochlorination by tetrachlorethylene (Ordoñez, 2001), as catalytic converter (exhaust fumes), more specifically in methane catalytic incineration (Paredes, 2004), removing sulphur from waste gas (Khalafalla & Haas, 1972), selective catalytic reduction of nitrogen dioxide (Lamonier, 1995), removing hydrogenchloride, carbon monoxide and dioxins (Hosoda et al., 1995) and recently also in ammonia decomposition (Ng et al., 2007). Other possibilities of catalytic use of red mud are described in literary review (Sushil & Batra, 2008).

Considering the size of specific surface of waste mud, its use as cheap adsorptive agent gives us a wide scale of possibilities in different areas. When preparing adsorbents from waste mud, it is necessary to realize firstly neutralization, washing, drying or, if necessary, another activation. It is important to note that, when there is variety of waste mud composition in dependence on processing technology (content of metals, radioactive elements, as well as organic substances – see chapter 2), it is required to perform leaching tests to manage the risk of possible contamination, e. g. in water treatment technology. Applications of red mud as adsorbent were presented (Wang et al., 2008) in adsorption of water solutions anions (PO_4^{3-}, F^-, NO_3^-), trace elements cations – metals and metalloids (Cu^{2+}, Pb^{2+}, Cd^{2+}, Zn^{2+}, Ni^{2+}, Cr^{6+}, As^{3+} and As^{5+}), pigments (rhodamine B, Congo red, methylene blue), organic substances (phenols, chlorinated phenols), but also in adsorption of toxic gases in emissions cleaning (H_2S, SO_2, NO_x). Titanium extraction by diluted sulphuric acid under atmospheric conditions was carried out without any previous treating of red mud, while following interactions of leaching process were monitored: acid concentration, temperature and ratio of solid and liquid part on which efficiency of titanium separation depended (recovery/yield of 64,5 % was achieved) (Smirnov & Molchanova, 1997). By the adsorption on red mud, there were described: separation of sulphuric compounds (cyklohexantiol) from oil (Singh et al., 1993), separation of phosphoric compounds (Mohanty et al., 2004) or removing bacteria and virus (*Escherichia coli, Salmonela adelaide* and poliovirus-1) by waste water filtration through sand filter mixed with red mud (Ho et al., 1991).

To prepare gypsum waste mud is elutriated by sulphuric acid or by mixture of H_2SO_4, HNO_3, KNO_3, soil microorganisms (*Deuteromyces*) and water in weight ratio of 10:3:14:72, with pH value < 1, still mixed for 100 - 200 hours. Created solid phase, after separation by filtration and drying, provides gypsum and potassium aluminium sulphate is obtained from filtrate after crystallisation. After crystallisation it is possible to thicken mother liquor by evaporation. Red pigment is gained from originated solid amorphous phase with majority share of hematite after drying, followed by calcination at the temperature of 600 °C and treating particles size by ultrafine milling under 5 µm (Kušnierová et al., 2001).

3.5 Agriculture and soil remediation

Extensive laboratory experiments were carried out in Australia (Summers et al., 2001). They dealt with red mud utilization as additive into soil to improve its properties during 10 years, when reduced loss of nutrients from soil into water was found out and agriculture production was increased (van Beers et al., 2009). Contaminated soil remediation by agents based on red mud was described. In the remediation there is adsorption of toxic substances and the structure and texture of soil improve. A series of applications was characterized in processing of acid mine drainage at sludge beds, which are old environmental loads and where application of alkaline red mud decreases soil acidity and, due to its excellent sorption properties, it binds a lot of toxic metals, e. g. Cd, As, Cu, Pb, Zn, Cr etc. (Gray et al., 2006; Bertocchi et al., 2006). Except for pH value increase, addition of red mud into soil shifts chemisorption of metal ions to Fe-oxide form. Consequently acidic extractability of these metals decreases and their fixation in red mud is ensured (Lombi et al., 2002). Synergistic effect of red mud additives to compost compounds, which provide for organic carbon increase in created soil, was studied, too (Brunori et al., 2005). In this soil, there was monitored sequential extractability of several fractions (poorly adsorbed, reducible, oxidizable and the rest) by using different extractants (CH_3COOH, $NH_2OH.HCl$, H_2O_2 + CH_3COONH_4 and compound of HCl, $HClO_4$ and HF). After addition of red mud into soils contaminated by heavy metals, decreasing of metal mobility was confirmed (the most significant leachability reduction was in case of Mn, Zn and Ni). This makes reusing of red mud in big quantities very promising (Maddocks et al., 2004).

Absorption ability of red mud to remove pollutants has to be supported by activation of fresh mud from refineries. Different activation methods and their effect on physiochemical properties are summarized in review (Liu, Y. et al., 2011).

3.6 Other utilization

Bauxite waste mud can be used in mining industry in gold extraction by cyanide method (Browner, (1995), where it is a very efficient modifier of pH value that has to be kept high in the process, while gold, possibly adsorbed on red mud, together with gold in ore, is separated by gravitational sedimentation. Recently a new method of production of shielding material against X-ray radiation using red mud and barium compounds has been developed (Amritphale et al., (2007). In comparison with traditional lead protective screen, needed thickness to retain the same energy was significantly lower, while other mineralogical (X-ray diffraction) and mechanical properties (pressure force and impact force) of new shielding material were studied, too.

Effective passivation of steel surface by using of red mud as anticorrosive agent (Collazo et al., 2005) before subsequent modifications (grinding, dipping, degreasing, painting etc.) was described. It was found out that, after immersing steel into decanting suspension of red mud and continuously mixed during 24 hours, depassivation runs at lower value of Cl^-/OH^- ratio and at lower pH value than untreated samples. The results of X-ray diffraction analysis confirmed that red mud does not create continuous layer on passivized steel surfaces, but it creates discrete distribution of individual elements formed mainly by oxy-hydroxides of Fe and Al. The development of electrode potentials on steel surfaces using electrochemical impedance spectroscopy was studied. It was studied in dependence on bating length in red mud bath, where, in comparison with untreated samples, significant potential of current density was reached (Collazo et al., 2007).

Hamdy and Williams (2001) studied bacterial amelioration of bauxite waste mud. They describe isolation of 150 bacterial cultures (the most significant representatives were *Bacillus, Lactobacillus, Leuconostoc, Micrococcus, Staphylococcus, Pseudomonas, Flavobacterium* and *Enterobacter*). After addition of nutrients and possible chemical modification the microorganisms were able to grow until the concentration of 10^9 $CFU.g^{-1}$ of waste mud, while organic acids produced by metabolism reduced pH value from 13 to 7. Aluminium recovery using *Penicillium simplicissimum* (Ward & Koch, 1996), calcium and iron using *Bacillus polymyxa* (Anand et al., 1996) were characterized by biological leaching. By bacterial metabolism, which runs by mechanism of oxidoreduction reactions using not only carbon but also sulphur and metals, it comes up to the metal nucleation at specific places of the cell surface of bacteria. Consequently, under appropriate conditions (especially pH control, aeration and nutrients addition), large amounts of bacterial clusters can catalyse secondary minerals formation (2005), which can be separated from original mixture.

4. Studies of environmental compatibility of waste mud

4.1 Leachability

The most important dangerous property of waste mud from Al_2O_3 production is high pH value because of the presence of large amount of NaOH in the leaching process that is used redundantly and it causes causticity or corrosivity of this waste. For this reason raw waste mud is usually neutralized before its placement to sludge bed. Most often it is neutralized by sea water or by other technologies.

Toxic effect of dangerous waste depends on biological accessibility of toxic agents, which is related to solubility in water medium, i. e. leachability. Recently several research of leachability of waste mud from Al_2O_3 production has been performed. Using extraction test TCLP (Toxicity Characteristic Leaching Procedure Test) at pH value of 3,0 Indian authors (Singh & Singh, 2002) studied metals extraction from red mud (mainly hexavalent Cr and Fe), where very low leachability was found out. The possibilities of Cr^{6+} leachability decrease from red mud were studied because of red mud use as a component of building materials and its stabilization by Portland cement in ratio of 3:1 was proposed. Metals leachability from Bauxsol™ was studied, too. Bauxsol™ presents red mud neutralized by sea water and in practice it is used as adsorption material. Detected leachability under different conditions did not exceed limiting values for heavy metals content in waste water (McConchie et al., 1996).

Besides red mud, leachabitility of heavy metals from different mixtures with red mud was observed. Red mud is added into these mixtures especially as effective adsorbent, e. g. in different soils, composts or in building materials. Noteworthy Italian kinetic study concerns leachability of trace elements using sequential extraction with red mud added to contaminated mining soil, where leachability of some metals, present in large amounts, e. g. Mn, Zn a Ni (Brunori et al., 2005a), was significantly decreased. In another study of Italian authors (Brunori et al., 2005b) leachability of differently modified red mud (neutralization, washing) was observed according to extraction test by deionised water in ratio of L/S = 5 and eightfold repetition of extraction in given time periods – 2, 8, 24, 48, 72, 102, 168 and 384 hours. Values of red mud leachability without washing and after washing by water are in Table 4.

Parameters	Unit	Unwashed treated red mud	Washed treated red mud	Threshold value*
Arsenic	µg.l⁻¹	45 ± 4	24 ± 3	50
Beryllium	µg.l⁻¹	< 0.5	< 0.5	5
Cadmium	µgl⁻¹	1.0 ± 0.1	0.6 ± 0.1	5
Cobalt	µg.l⁻¹	< 0.5	< 0.5	50
Chromium	µg.l⁻¹	16 ± 2	6 ± 1	50
Copper	µg.l⁻¹	51 ± 5	2,7 ± 0,5	100
Nickel	µg.l⁻¹	11 ± 1	1,5 ± 0,3	100
Lead	µg.l⁻¹	1.7 ± 0.2	5 ± 1	50
Vanadium	µg.l⁻¹	555 ± 50	485 ± 70	50
Zinc	µg.l⁻¹	< 50	< 50	3 000
Chloride	mg.l⁻¹	4880 ± 450	128 ± 25	200
Fluoride	mg.l⁻¹	32 ± 3	16 ± 3	1.5
Nitrates	mg.l⁻¹	2.4 ± 0.2	n.d.	50
Sulphates	mg.l⁻¹	1140 ± 100	530 ± 100	250
pH	range	6.4 – 10.5	8.2 – 10.2	5.5 – 12

*For the Italian leaching test

Table 4. Leachability of neutralised red mud without washing and after washing by deionised water (Brunori et al., 2005b).

Measured above the limit values of some anions were probably caused by processing of red mud by seawater. After washing of such processed mud by deionised water, significant decrease of anion content was reached, in case of chlorides there was 30-fold decreasing. Despite washing, content of sulphates and fluorides was still above limits. From monitored metals vanadium is important, because its concentration is 10-fold above the limit (according to Italian legislation twice above the limit). This high concentration is ascribed by authors to high mobility of vanadium in neutral and alkali environment and also to high content of vanadium in the red mud.

According to EPA report (U.S. EPA, 1990) about special waste from minerals treating for Al_2O_3 production, except for selenium and arsenic, all compounds monitored according to EPA rules were at least two orders below the limits. From 18 indicators monitored in elutriate from red mud, As and Se concentrations were three time above screening criteria, what means (in precondition of elutriate diluting by ground water) at ingestion a cancerogenity risk for people. Despite strict screening criteria of EPA for elutriate, the Report (U.S. EPA, 1990) ascertains that toxicity limits were not exceeded in all monitored cases.

As described by Friesl (Friesl et al., 2003), after red mud adding to sand soils, significant decrease of heavy metals leachability was observed, e.g. Cd by 70 %, Zn by 89 % and Ni by 74 %. Simultaneously authors monitored catching of these metals by plants, what lead to decrease of metal content by 38 – 87 %. However, if there is added red mud into the soil exceeded 5 wt.%, content of some toxic metals (As, Cr and V) is increased too, and changes in bioavailability of the metals are also discussed. For instance, after increasing of total Cu content after red mud adding, probably because of decrease of Cu bioavailability, decreasing of phytotoxicity occurred what is confirmed by higher production of corn biomass. There were also found out concentrations of other monitored metals: arsenic 5 mg.kg^{-1}, chrome 20 mg.kg^{-1} and vanadium 5 mg.kg^{-1}. After leaching these amounts cause overcoming of limits and therefore adding of red mud into the soil to enhance of soil properties must be controlled (Friesl et al., 2004). Effect of pH was studied to soils enhanced by addition of red mud, lime and beringite and was confirmed leachability decreasing of Cu, Cd and Zn (Lombi et al., 2003). The leachability is low in wide range of soil pH despite reacidification of soil.

4.2 Toxicity tests and ecotoxicological biotests

Ecotoxicologic biotests and chemical analyses are the most important methods of ecotoxicological detection systems for elutriates from waste. These methods can directly assess detrimental effects on environment regardless off composition and character of tested waste. Significance of these methods consists in identification and assessment of risk mainly from the point of view of migration of contaminants from soil into ground water. The biotests are realized at producer-consumer-decomposer level and its combination can improve or degrade the result of extraction methods. Choosing of appropriate leachability method and detection system are the biggest sources of variability in assessment of leachates from waste. At classification of waste into the hazardousness categories, an economic factor plays important role. Therefore the biggest abuse of factors influencing results of tests at choosing of above mentioned systems from waste producers is assumed. For the kind of waste with extremely high pH, different methods of extraction had to been ordained. These methods were optimised according to the experiments.

Apart from leachability tests, important mainly in landfilling and assessment of environment contamination – contamination of ground water, surface water near sludge bed, ecotoxicity tests of sludge are also important. The tests enable to assess appropriate way of using in above mentioned spheres. Ecotoxicity tests of red mud have been described in literature only recently and just for small number of applications.

In the environmental compatibility study of red mud from Italian authors (Brunori et al., 2005b) three kinds of standard ecotoxicologic tests were realized. Red mud was neutralised by sea water and final pH of processed mud was about 7. Acute toxicity was tested by sea luminescent bacteria *Vibrio fischeri* with biotest Microtox™. Concentration from 0.1 to 2.0 % of red mud in sea water was measured. Obtained value of bioluminescence produced by the bacteria did not reach the detection limit of the method (20 %), what refers to negative environmental effect of the suspension. Similar result was measured by the test with *Dunaliella tertiolecta* according to American Society for Testing Materials (ASTM) methodology. No significant differences between blank and water extract of neutralised red mud at above mentioned concentrations were measured by ASTM test of embryonic toxicity in pluteal phase of sea gastropod *Sphaeroechinus granularis* after 72-hour incubation. Only in paper of Pagano (Pagano et al., 2002), toxicity of raw red mud on sea urchin embryos was detected because of high alkalinity (pH 12). In various samples from sludge bed of 4 European aluminium works (Turkey, France, Greece and Italy), larval retardation, malformation, development malfunctions and early embryonic mortality were studied. Significant sperm toxicity and influence to offspring quality was confirmed.

Acute toxic effect of adsorbent Bauxsol™, which is produced from red mud after neutralisation and processing with sea water in Australia (Corp. Virotec Global Solutions of Gold Coast, Queensland, Australia), was studied by earthworms (*Eisenia foetida*) (Maddocks et al., 2005). Tested Bauxsol™ contained more than 6,450 mg of adsorbed metals per 1 kg of adsorbent. The Bauxsol™ was mixed with dung of cows into various ratios. Final concentrations of the Bauxsol™ were 100 %, 90 %, 80 %, 60 %, 40 % and 20 %. Earthworms were exposed to the mixtures for 28 days. In all samples earthworms exhibited good mobility and no mortality. The highest bioaccumulation of metals in earthworms was found out at 20 and 40 % of Bauxsol™ in dung of cows. Despite it, calculated bioaccumulation factors (BAF) were lower than published threshold values of toxicity causing mortality or published values for middle polluted soils. By sequence extractive analysis of mixture 20 % Bauxsol™ in dung of cows it was found out, that more than 95 % of metals were bound in Fe/Mn oxide fraction and after exposing by earthworms, change in composition of metal fraction Cd/Cr and Fe/Mn happened. Mentioned results indicate that metals adsorbed in Bauxsol™ are not accessible for earthworms and extractive methods are usable for studying of toxicological characteristics.

Genotoxic properties of red mud have been studied by Oreščanin (Oreščanin et al., 2003). Tested samples exhibited no cytotoxic and mutagenic activity at two bacterial strains *Salmonella typhimurium* (TA98 a TA100), which were studied at presence and also at absence of metabolic activation. The authors used sequence extractive analysis of red mud. No toxic effect of new coagulant prepared from red mud was confirmed before starting commercial production of the coagulant. Reusing and regeneration of waste mud were described as very promising.

Relatively wide-ranging and long term ecotoxicity tests of red mud were realized also in France (Ribera & Saint-Denis, 2002) in connection with depositing of waste from two aluminium works near Marseilles on the sea bed of Lion bay. In decade 1997 – 2007 all ecotoxicity tests were negative - Microtox® by *Vibrio fisheri*, chronic toxicity tests by sea-urchins, gastropods and oysters, Ammes test of genotoxicity and acute tests by European seabass *Dicentrachtus labrax* - 152 tests total from 19 sampling places, except for two contact tests by larval phase of sea-urchin, where relatively low number of development anomalies were observed (< 39 %). In the study (Dauvin, 2010), besides ecotoxicologic aspects there is described also effect of deposited red mud on the sea bed relief, on mezzo fauna and macro fauna in connection with dispersion of red mud in the sea environment. There were also studied changes in macro benthos of deep sea communities and risks of consummation of sea animals which were in contact with waste sludge from Al_2O_3 production. Last mentioned study was carried out at risk part of human population (women, children) and no cumulative risk for majority elements (Al, Fe, Cr, Pb, Cu, Mn, V and Zn) was confirmed.

According to EPA report (U.S. EPA, 1982; U.S EPA, 2008) red mud after neutralisation is not classified as hazardous waste (Wang et al., 2008), because in four tested hazardous properties (corrosivity, reactivity, ignitability and toxicity according to TCLP (Toxicity Characteristic Leaching Procedure Test) it does not fulfil criteria for such classification. Performed ecotoxicity tests indicate that neutralised red mud can be widely reused not only as a building material, raw material for metal production in metallurgic industry or in glass production. Because of large surface, red mud after activation becomes excellent adsorbent and coagulant which can be used in remediation of soil in agriculture, mining industry - neutralisation of Acid Mine Drainage (Paradis et al., 2006), in removal of toxic metals in waste treatment plants, in catalysis in chemical industry etc. At many above mentioned applications of red mud, its benefit effect on environment was confirmed by experiments (e.g. improvement of soil properties after adding of red mud into unproductive (thin) clay or sand soils, or as a catcher of toxic inorganic and organic substances not only from soil, but also from water and air, as a gas cleaner. While heavy metals remain adsorbed and leachate contains safety levels of the metals even in low pH – acid rains). Despite huge amount of mentioned applications of the mud reusing, because of low economic profit rate the mud in many cases ends up at sludge beds.

5. Conclusion

From this literature survey it is obvious that there is extensive effort to reusing of waste mud from aluminium production. Nowadays, patent databases register worldwide more than 1 500 patents related to red mud and its reusing in different fields mentioned in this paper. On the other hand, it is claimed in general that utility of this patent is very low because of economic reasons. A lot of interesting references can be found at web sites directly dedicated to red mud reusing (Red mud project, 2011). Despite the fact that red mud is classified in some countries as hazardous waste because of strong alkalinity (according to OECD classification to Yellow list of waste under code GG 110-SRA), after neutralisation its effect on environment seems to be negligible from the toxicity point of view. A lot of leaching experiments were performed with waste mud with various extractive agents (so-called sequence extractive analysis), mostly to find out toxic metals leachability

(Singh, I. B. & Singh, D. R., 2002; Friesl, et al., 2003). In general limits for waste water were not exceeded in all these experiments. Reusing of waste sludge in building industry to produce final articles requires also performing of acute and chronic ecotoxicity tests by relevant biologic species. Results obtained up to the present (Maddocks et al., 2005; Oreščanin et al., 2003; Brunori et al., 2005b) have not confirmed toxic effects of tested articles or processed red mud on tested organisms (algae, worms, marine gastropods, etc.), meanwhile unprocessed red mud without neutralisation exhibits toxic effect on water organisms (Pagano et al., 2002).

In spite of intensive research of reusing waste mud from Al_2O_3 production, majority of the mud ends up on terrestrial deposits. In seaside areas, where the distance to the sea is not too long, efforts to deposit red mud into the sea after neutralisation by sea water occur (from 84 world producers of Al_2O_3, deposits of the mud are on the sea bed only in 7 cases) (Agrawal et al., 2004), what results in contradictory reactions of environmentalists. In many countries the problem of sludge bed is solved by neutralisation of alkaline waste water, building of underground sealing wall from bentonite to stop leaching the waste water into surrounding ground water and subsequent recultivation of the sludge bed. One of the major limitations of the successful exploitation of bauxite residue is large transport cost necessary to transfer of waste mud from its disposal sites to the point of application.

6. Acknowledgement

This contribution is the result of the project implementation: Centre of excellence for white-green biotechnology, ITMS 26220120054, supported by the Research & Development Operational Programme funded by the ERDF (50 %) and by the Ministry of Education of the Slovak Republic KEGA No. 007TUZVO-4/2011, KEGA No. 011TUZVO-4/2011 and APVV SK-CZ-0139-11 (50 %).

7. References

Agatzini-Leonardou, S., Oustadakis, P., Tsakiridis, P. E. & Markopoulos, Ch. (2008). Titanium leaching from red mud by diluted sulfuric acid at atmospheric pressure. *Journal of Hazardous Materials.* Vol.157, 2008, pp. 579-586, ISSN 0304-3894

Agrawal, A.; Sahu, K. K. &, Pandey, B. D. (2004). Solid waste management in non-ferrous industries in India. *Resources Conservation and recycling,* Vol.42, 2004, pp. 99-120, ISSN 0921-3449

Álvarez, J.; Ordóñez, S.; Rosal, R.; Sastre, H. & Díez, F. V. (1999). A new method for enhancing the performance of red mud as a hydrogenation catalyst. *Applied Catalysis A: General.* Vol.180, No.1-2, (1999). pp. 399-409

Amritphale, S. S. & Patel M. (1987). Utilization of red mud, fly ash for manufacturing bricks with pyrophyllite. *Silicates Industriels.* 1987; Vol.52, No.3-4, pp.31-35.

Amritphale, S. S.; Anshul, A.; Chandra, N. & Ramakrishnan, N.(2007). A novel process for making radiopaque materials using bauxite - Red mud. *Journal of the European Ceramic Society,* Vol.27, No.4, 2007, pp. 1945-1951, ISSN 0955-2219

Anand, P.; Modak, J. M. & Natarajan, K. A. 1996. Biobeneficiation of bauxite using Bacillus polymyxa: calcium and iron removal. *International Journal of Mineral Processing* Vol.48, No:1-2, november 1996, pp. 51-60, ISSN 0301-7516

Andrejcak, M. & Soucy, G. (2004). Patent Review of Red Mud Treatment – Product of Bayer Process. *Acta Metallurgica Slovaca*, Vol.10, 2004, pp. 347-368

Akinci, A. &; Artir, R. (2008). Characterization of trace elements and radionuclides and their risk assessment in red mud. *Materials Characterization*, Vol.59, No.4, April 2008, pp. 417-421

Bardossy, G.; Aleva, G.J.J. (1990). *Lateritic Bauxites. Developments in Economic Geology Vol.27.* Elsevier Sci. Publ. 1990, 624 pp.

Bertocchi, A. F.; Ghiani, M.; Peretti, R. & Zucca, A. (2006) Red mud and fly ash for mine sites contaminated with As, Cd, Cu, Pb and Zn. *Journal of Hazardous Materials* Vol.134, 2006, pp.112–119, ISSN 0304-3894

Bhatnagar, A.; Vilar, V.J.P.;. Botelho, C.M.S & Boaventura R.A.R. (2011). A review of the use of red mud as adsorbent for the removal of toxic pollutants from water and wastewater. *Environmental Technology*, Vol.32, No.3, 2011, pp. 231-249

Brunori, C.; Cremisini, C.; D'Annibale, L.; Massanisso, P. & Pinto, V. (2005a). A kinetic study of trace element leachability from abandoned-mine-polluted soil treated with SS-MSW compost and red mud. Comparison with results from sequential extraction. *Analytical and Bioanalytical Chemistry*, Vol.381, No. 7, 2005, pp 1347-1354, ISSN 1618-2642

Brunori, C.; Cremisini, C.; Massanisso, P.; Pinto, V. & Torricelli, L. (2005b). Reuse of a treated red mud bauxite waste: studies on environmental compatibility, *Journal of Hazardous Materials*, Vol.117, No.1, 2005, pp.55-63, ISSN 0304-3894

Browner, R. E. (1995). The use of bauxite waste mud in the treatment of Gold ores. *Hydrometallurgy*, Vol.37, No.3, April 1995, pp. 339-348, ISSN: 0304-386X

Buraev, M. I. & Kushnir, L. I. (1986). Facing tiles obtained from hydromica clays and red mud. *Kompleksnoe Ispol'zovanie Mineral'nogo Syr'ya*, Vol.7, 1986, pp. 66-69

Collazo, A.; Fernandez, D.; Izquierdo, M.; Nóvoa, X. R. & Pérez, C. (2005). *Progress in Organic Coatings.* Vol.52, 2005, pp. 351–358, ISSN 0300-9440

Collazo, A.; Izquierdo, M.; Nóvoa, X.R. & Pérez, C. (2007). Surface treatment of carbon steel substrates to prevent cathodic delamination. *Electrochimica Acta* Vol.52, No.27, 7513-7518, ISSN 0013-4686

Dass, A. & Malhotra, S.K. (1990). Lime-stabilized red mud bricks. *Materials and Structures* Vol.23, No.4, pp. 252-255 (1990).

Dauvin J. C. (2010). Towards an impact assessment of bauxite red mud waste on the knowledge of the structure and functions of bathyal ecosystems: The example of the Cassidaigne canyon (north-western Mediterranean Sea) *Marine Pollution Bulletin* Vol.60, No.2, 2010, pp. 197–206

Di San Filippo A. (1980). Riutilizzo del fango rosso. *Rendiconti del seminario della Facolt`a di Scienze dell'Universit`a di Cagliari*, Vol. L, No. 3–4, Cagliari, Italy, 1980

Dimas, D. D.; Giannopoulou, I. P. & Panias D. (2009). Utilization of alumina red mud for synthesis of inorganic polymeric materials. *Mineral processing and Extractive Metallurgy. Review*, Vol.30, No.3, 2009, pp. 211-239

Eamsiri, A.; Jackson, W.R.; Pratt, K. C.; Christov, V. & Marshall, M. (1992). Activated red mud as a catalyst for the hydrogenation of coals and of aromatic compounds. *Fuel*, Vol.71, No. 4, (1992). pp. 449-453

Ercag, E. & Apak, R. (1997). Furnace smelting and extractive metallurgy of red mud: Recovery of TiO₂, Al₂O₃ and pig iron. *Journal of Chemical Technology and Biotechnology*. Vol.70, November 1997, pp. 241-246, ISSN 1097-4660

Friesl, W.; Lombi, E.; Horak, O. & Wenzel, W.W. (2003). Immobilization of heavy metals in soils using inorganic amendments in a greenhouse study. *Journal of Plant Nutrition and Soil Science*, Vol.166, No.2, 2003, pp. 191-196, ISSN 1436-8730.

Friesl, W.; Horak, O. & Wenzel, W. W.: (2004). Immobilization of heavy metals in soils by the application of bauxite residues: pot experiments under field conditions *Journal of Plant Nutrition and Soil Science*, Vol.167, No.1, 2004, 54-59, ISSN: 1436-8730

Gheorghita, M. & Sirbu, E. (2009). New methods to obatin high-purity Gallium. *Metalurgia International*, Vol.14, pp.65-68

Gray, C. W.; Dunham, S. J.; Dennis, P. G.; Zhao, F. J. & McGrath, S. P. (2006). Field evaluation of in situ remediation of a heavy metal contaminated soil using lime and red-mud. *Environmental Pollution*, Vol.142, No.3, Aug 2006, pp. 530-539, ISSN 0269-7491

Hamdy, M. K. & Williams, F. S.: J. (2001). Bacterial amelioration of bauxite residue waste of industrial alumina plants. *Journal of Industrial Microbiology & Biotechnology*, Vol.27, No.4, Oct 2001, pp. 228–233, ISSN 1367-5435

Hetherington, L.E.; Brown, T.J.; Benham, A.J.; Lusty, P.A.J. & Idoine, N. E. (2007). *World Mineral Production: 2001 - 2005*. British Geological Survey 2007

Hind, A. R.; Bhargava, S. K. & Grocott S. C. (1999). The surface chemistry of Bayer process solids: a review *Colloids and Surfaces A: Physicochemical and Engineering Aspects* Vol. 146, pp. 359- 374 ISSN 0927-7757 (1999).

Ho G. E., Gibbs R. A., Mathew K. (1991). Bacteria and virus removal from secondary effluent in sand and red mud columns. *Water Science and Technology*. Vol.23, 1991, pp. 261-270, ISSN 0273-1223

Hosoda, H.; Hirama, T. & Aoki, H. (1995). Simultaneous reduction techniques of nitrous and nitrogen oxides from fluidized-bed coal combustor. *Kagaku Kogaku Ronbunshu*, Vol.21, No.1, 1995, pp. 74-82, ISSN 0386-216X

Idaho State University (2009). Radiation Information Network's. Radioactivity in Nature. (10.9.2011) Available from http://www.physics.isu.edu/radinf/natural.htm

Kane, J. (1979). US Patent, 4146573, Derwent 79-29434 (1979).

Khalafalla S. E. & Haas L. A. (1972). The role of metallic component in the iron-alumina bifunctional catalyst for reduction of SO₂ with CO. *Journal of Catalysis*. Vol.24, No. 1, January 1972, pp. 121-129, ISSN 00219517

Kohno, K.; Sugimoto, A. & Kashiwai T. (1998). High-strength concrete containing finely ground silica and red mud. *Semento Gijutsu Nenpo 42*, 136 (1998); *Chemical Abstracts 111*, 083124 (1998)

Klauber, C., Gräfe, M. & Power, G. (2009). Review of Bauxite Residue "Re-use" Options. CSIRO Document DMR-3609, National Research Flagships, Light Metals, Australia, May 2009, pp. 1-79, 20.10.2011 Available from:
http://www.asiapacificpartnership.org/pdf/Projects/Aluminium/Review%20of%20Bauxite%20Residue%20Re-use%20Options_Aug09_sec.pdf

Klein, M. (1998). German Patent, 3633413, *Chemical Abstracts 108*, 191683 (1998)

Kušnierová, M.; Kafka, R. & Vašková, H. (2001). Int. Cl: B09B 3/00 Slovak patent, 281359 Úrad priemyselného vlastníctva SR, Banská Bystrica, 2001

Lamonier, J. F.; Leclerco, G.; Dufour, M. & Leclercq, L. (1995). Utilization of red mud. Catalytic properties in selective reduction of nitric oxide by ammonia. Recents Progres en Genie des Procedes, (43, Boues Industrielles: Traitements, Valorisation), pp. 31-36

Lamerant, J. M. (2000). (Aluminum Pechiney, Courbevoie, France) Int. Cl. B01D 15/04, United States Patent, 6110377 (2000)

Li, B.; Xu, Y. & Choi, J. (1996). Applying Machine Learning Techniques, Proceedings of ASME 2010 4th International Conference on Energy Sustainability, Phoenix, Arizona, USA, May 17-22, 2010, pp. 14-17, ISBN 842-6508-23-3

Li, L. Y. (2001). A study of iron mineral transformation to reduce red mud tailings. Waste Management Vol.21, pp. 525-534, ISSN 0956-053X

Liu, W.; Yang, J. & Xiao, B. (2009). Application of Bayer red mud for iron recovery and building material production from aluminosilicate residues. Journal of Hazardous Materials, Vol.161, pp.474-478. ISSN 03043894

Liu, Y.; Naidu, R. & Ming, M. (2011). Red mud as an amendment for pollutants in solid and liquid phases. Geoderma Vol.163, No.1-2, (15 June 2011), pp 1-12, ISSN 0016-7061

Lombi, E.; Zhao, F. J.; Zhang, G.; Sun, B.; Fitz, W.; Zhang, H. & McGrath, S. P. (2002) In situ fixation of metals in soils using bauxite residue: chemical assessment. Environmental Pollution, Vol.118, No.3, 2002, pp. 435-443, ISSN 0269-7491

Lombi, E.; Hamon, R. E.; McGrath, S. P. & McLaughlin, M. J. (2003). Lability of Cd, Cu, and Zn in polluted soils treated with lime, beringite, and red mud and identification of a non-labile colloidal fraction of metals using istopic techniques. Environmental Science Technology, Vol.37, No.5, Mar 2003 , pp. 979-984, ISSN 0013-936X

Maddocks, G.; Lin, C. & McConchie, D. (2004). Effects of Bauxsol™ and biosolids on soil conditions of acid-generating mine spoil for plant growth. Environmental Pollution Vol.127, No.2, Jan 2004, pp.157-167, ISSN 0269-7491

Maddocks, G.; Reichelt-Brushett, A.; McConchie, D. & Vangronsveld, J. (2005). Bioaccumulation of metals in Eisenia fetida after exposure to a metal-loaded bauxsol™ reagent. Environmental Toxicology and Chemistry, Vol. 24, No.3, March 2005, pp. 554–563, ISSN 1552-8618

McConchie, D.; Saenger, P. & Fawkes, R. (1996). An environmental assessment of the use of seawater to neutralise bauxite refinery wastes. In: Ramachandran, V., Nesbitt, C.C. (Eds.), Proceedings of 2nd International Symposium on Extraction and Processing for the Treatment and Minimization of Wastes. Minerals, Metals and Materials Society, Scottsdale, AZ, 1996, pp. 407–416

Mohanty, S.; Pradhan, J.; Das, S. N. & Thakur, R. S. (2004). Removal of phosphorus from aqueous solution using alumized red mud. International Journal of Environmental Studies, Vol.61, 2004, pp. 687–697, ISSN 0020-7233

Ng, P. F.; Li, L.; Wang, S.; Zhu, Z.; Lu, G. & Yan, Z. (2007). Catalytic Ammonia Decomposition over Industrial-Waste-Supported Ru Catalysts. Environmental Science Technology, 2007, Vol.41, pp. 3758-3762, ISSN 0013-936X

Ogura, H. (1978) Japanese patent, 7182527; Chem. Abstr. 089, 079289 (1978).

Ochsenkühn-Petropoulou, M. T.; Lyberopulu, T. & Parissakis H. (1995). Selective Separation and Determination od Scandium from Yttrium and Lanthanides in Red Mud by a Combined Ion Exchange/Solvent Extraction Method. Analytica Chimica Acta, 1995, Vol. 315, No1-2, pp. 231-237, ISSN 0003-2670

Ochsenkühn-Petropoulou, M. T.; Lyberopulu, T., Ochsenkühn, K. M. & Parissakis H. (1995). Recovery of Lanthanides and Yttrium from from red Mud by Selective Leaching. *Analytica Chimica Acta,* 1996, Vol. 319, No1-2, pp. 249-254, ISSN 0003-2670

Ochsenkühn-Petropoulou, M. T.; Hatzilyberis, K. S.; Mendrinous L. N. & Salmas, C. E. (2002). Pilot-plant investigation of the leaching process for the recovery of scandium from red mud. *Industrial & Engineering Chemistry Research,* Vol.41, No.23, November 2002, pp. 5794-5801

Ordoñez, S.; Sastre, H. & Diez, F. V. (2001). Catalytic hydrodechlorination of tetrachloroethylene over red mud, *Journal of Hazardous Materials,* Vol.B81, (2001). 103-114, ISSN 0304-3894

Orešcanin, V.; Durgo, K.; Franekic-Colic, J.; Nad, K. & Valkovic, V. (2003). Physical, chemical, and genotoxic properties of waste mud byproduct of waste water treatment. *Journal of Trace and Microprobe Techniques,* Vol.21, No.1, 2003, pp. 123-132, ISSN 0733-4680

Pagano, G.; Meric, S.; De Biase, A.; Iaccarino, M.; Petruzzelli, D.; Tunay, O. & Warnau, M. (2002). Toxicity of bauxite manufacturing by-products in sea urchin embryos, *Ecotoxicology and Environmental Safety,* Vol. 51, 2002, pp. 28–34, ISSN 0147-6513

Paradis M., Duchesne J., Lamontagne A., Isabel D. (2006). Using red mud bauxite for the neutralization of acid mine tailings: a column leaching test. *Canadian Geotechnical Journal,* Vol.43, No.11, 2006, pp. 1167-1179, ISSN 0008-3674

Paramguru, R.K.; Rath, P.C. & Misra, V.N. (2005) Trends in red mud utilization - A review *Mineral Processing and Extractive Metallurgy Rev. 26,* 1 , ISSN 0371-9553

Paredes, J. R.; Ordóñez, S.; Vega, A. & Díez, F. V. (2004). Catalytic combustion of methane over red mud-based catalysts. *Applied Catalysis B: Environmental,* Vol.47, 2004, pp. 37-45, ISSN 0926-3373

Pratt, K. C. & Christoverson, V. (1982). Hydrogenation of a model hydrogen-donor system using activated red mud catalyst. *Fuel,* Vol.61, No. 5, 1982, pp.460-462, ISSN 0016-2361

Piga, L.; Pochetti, F. & Stoppa L. (1995). Application of thermal-analysis techniques to a sample of red mud – a by-product of the Bayer process for magnetich separation. *Thermochimica Acta,* Vol.254, 1995, pp. 337-345, ISSN 0040-6031

Qiu, G.Z; Liu, Y.K; Jiang, T; Hu, Y.H. & Mei, X.G. (1996). Influence of additives on slag-iron separation during direct reduction of coal-base high-iron-content red mud. *Transactions of nonferrous metals Society of China.* Vol.6, No.2, JUN 1996, pp. 1-7

Red mud project, (2011). 10.9.2011, Available from <http://www.redmud.org/home.html>

Ribera, D. & Saint-Denis, M. (2002). Evaluation des dangers et gestion des risques. Quelques perspectives en e cotoxicologie animale. *Details - Bulletin de la Société zoologique de France,* Vol.127, 2002, pp. 329, ISSN: 0037-962X

Satish, Ch. (1997). *Waste Material Used in Concrete Manufacturing.* In: Division of Concrete Structures, Göteborg: Chalmers University of Technology, Sweden, 1997. p. 290, ISBN 0-8155-1393-3

Sglavo, V. M.; Campostrini, R.; Maurina, S.; Carturan, G.; Monagheddu, M.; Budroni, G. & Cocco, G. (2000a). *Journal of the European Ceramic Society* Vol.20, No.3, 2000, pp. 235-244, ISSN 0955-2219

Sglavo, V. M.; Campostrini, R.; Maurina, S.; Carturan, G.; Monagheddu, M.; Budroni, G. & Cocco, G. (2000b). Bauxite 'red mud' in the ceramic industry. Part 2: Production of clay-based ceramics. *Journal of the European Ceramic Society* (2000).Vol.20, No.3, pp. 245-252, ISSN 0955-2219

Singh, I. B. & Singh, D. R. (2002). Cr(VI) removal in acidic aqueous solution using iron-bearing industrial solid. *Environmental Technology*, Vol.23, No.1, January 2002, pp. 85-95, ISSN 0959-3330

Singh, A. P.; Singh, P. C. & Singh, V. N. (1993). Cyclohexanethiol separation from kerosene oil by red mud. *Journal of Chemical Technology and Biotechnology*, Vol.56, No. 2, 1993, pp. 167-174 ISSN 0268-2575

Singh, M.; Upadhayay, S. N. & Prasad, P. M. (1996). Preparation of Special Cements from Red Mud, *Waste Management*, Vol.16, No.8, (1996). pp.665-670, ISSN 0956-053X

Snars K. & Gilkes R.J. (2009). Evaluation of bauxite residues (red muds) of different origins for environmental applications. *Applied Clay Science*, Vol.46, No.1, 2009, pp. 13–20, ISSN 0169-1317

Smirnov, D. I. & Molchanova, T. V. (1997). The investigation of sulphuric acid sorption recovery of scandium and uranium from the red mud of alumina production. *Hydrometallurgy*, Vol.45 , July 1997, pp. 249-259, ISSN 0304-386X

2005. Geomicrobiology in contemporary natural systems: Implications for Economic Geology. *Economic Geology* Vol.100, pp. 1067-1084

Sushil, S. & Batra, V. S. (2008). Catalytic applications of red mud, an aluminium industry waste: A review. *Applied Catalysis B: Environmental*, Vol.81, No. 1-2, May 2008, pp. 64-77, ISSN 0926-3373

Summers, R. N.; Bolland, M. D. A. & Clarke, M. F. (2001). Effect of application of bauxite residue (red mud) to very sandy soils on subterranean clover yield and P response. *Australian Journal of Soil Research*. Vol.39, No.5, 2001, pp. 979 – 990, ISSN 1838-675X

Thakur, R. S. & Sant, B. R. (1983). Utilization of red mud. 2. Recovery of alkali, iron, aluminium, titanium, and other constituents and the pollution problems. *Journal of Scientific and Industrial Research*, Vol.42, No.8, pp. 456-469. ISSN 0022-4456

Tsakiridis, P. E.; Agatzini-Leonardou, S. & Oustadakis P. (2004). Red mud addition in the raw meal for the production of Portland cement clinker. *Journal of Hazardous Materials*, Vol.*116*, No.1-2, 2004, pp. 103-110, ISSN 0304-3894

U.S. EPA (1982). U.S. Environmental Protection Agency: Emissions of Naturally Occurring Radioactivity from Aluminum and Copper Facilities, Office of Radiation Programs, Las Vegas Facility, NV, p. 8

U.S. EPA (1990). U.S. Environmental Protection Agency: Aluminum Production from Report to Congress on Special Wastes from Mineral Processing, Vol. II, Office of Solid Waste, July 1990, pp. 3-11

van Beers, D.; Bossilkov, A. & Lund, C. (2009). Development of large scale reuses of inorganic by-products in Australia: The case study of Kwinana, Western Australia. *Resources, Conservation and Recycling*, Vol.53, No.7, May 2009, pp. 365-378

Wagh, A. S. & Douse, V.E. (1991). Silicate bonded unsintered ceramics of Bayer process waste. *Journal of Material Research*, Vol.6, No. 5, 1991, pp. 1094–1102

Wang, S.; Ang, H. M. & Tadé, M. O. (2008): Novel applications of red mud as coagulant, adsorbent and catalyst for environmentally benign processes. *Chemosphere* Vol.72, No. 11, 2008, pp. 1621-1635, ISSN 0045-6535

Ward, S. C. & Koch, J. M. (1996). Biomass and nutrient distribution in a 15.5 year old forest growing on a rehabilitated bauxite mine. *Australian Journal of Ecology* (1996) Vol.21, 1996, pp. 309-315, ISSN 1442-9993

Yalcin, N. & Sevinc V. (2000). Utilization of bauxite waste in ceramic glazes. *Ceramics International* (2000), Vol.26, No.5, pp. 485-493, ISSN 0272-8842

Zhong, L.; Zhang, Y. & Zhang, Y. (2009). Extraction of alumina and sodium oxide from red mud by a mild hydro-chemical process. *Journal of Hazardous Materials*, Vol.172, No.2-3, 30 December 2009, pp. 1629-1634, ISSN 0304-3894

Part 2

Hydro-Electrometallurgical Processes

2

Methods for the Enhancement of Mass Transport at the Recovery of Metal Ions from Hydroelectrometallurgical Processes

Mirela Ioana Iorga, Raluca Pop,
Marius Constantin Mirica and Doru Buzatu
National Institute for Research & Development in Electrochemistry
and Condensed Matter – INCEMC – Timisoara,
Romania

1. Introduction

The electrochemical deposition of metals from aqueous solutions represents the basis of the hydroelectrometallurgical processes: the extraction of metals from ores (electroextraction or electrowinning) and their purification by electrolysis (electrorefining). Some of the metals that are obtained and refined from hydroelectrometallurgical processes are copper, nickel, zinc, cadmium, tin, lead, silver, gold and manganese. The hydroelectrometallurgical processes lead to the obtaining of technical pure metals and to an economic recovery of metals from poor ores. Electrodeposition may be also used as preconcentration technique for trace analyses.

Another important application of this process is represented by the treatment of the waste waters that contain metal ions. The classical depollution methods are no longer sufficient to the new, harsher legal provisions and they are also characterized by high costs. As a consequence, modern methods as the removal by electrodeposition of the pollutant metallic ions from effluents and from the processes flows have been developed. In these procedures, ions are deposited on the structure of a support electrode in order to be recycled or removed. It is thus prevented the dissipation of toxic metals in soil from the waste storage dumps from industrial processes.

It was necessary to establish the effects of pollutants that contain metal ions, which can act on short or long term, due to accumulation of toxic substances and their derivatives in the human body (Walsh and Reade, 1994). Regulations in this area have considered the nature of polluting materials, applicable limits of long- and short-term exposure, procedures/rules for handling and storage of toxic substances.

A credible recycling must be supported both in environmental and in economic terms (Emery et al., 2002; Zhang & Forssberg, 1998). According to the principles and to the strategic elements for a sustainable development, current trends are converging more and more towards the treatment of the waste waters at source. This approach of the problem is advantageous for substances that can be recycled in the process or that can be recovered in order to exploit them in other processes or as by-products (Iorga et al., 2007).

2. Sources of pollution with metal ions vs. the need for their recovery

Pollutants are present in certain concentrations in the environment as a result of human activity, having a significant negative impact on it. Due to the complexity and the high cost of the methods for their treatment, a detailed analysis in order to determine the best methods for their elimination is required.

The main sources of pollution with metal ions are the residues from industrial processes in the following areas:

- metallurgy (etching, polishing);
- mining (primary output of ores, mine waters);
- hydroelectrometallurgy (depleted electrolytes);
- electrochemistry (depleted electrochemical sources – batteries and accumulators, electroplating, electrodeposition, contaminated baths);
- chemistry (catalysts, chemical reagents, dyes);
- leather (tanning);
- photographic technique (fixing solution), etc.

It must be mentioned that in each industrial unit wastewater containing metal ions are obtained. It must be not forgotten the fact that all installations and objects made of metal are subjected both to erosion and especially to chemical corrosion and atmospheric agents (Kuhn, 1971; Walsh & Reade, 1994).

Metal recovery from recyclable waste is a less energy consuming process than getting the metal from raw mineral resources (also taking into account the increasing limitation of resources). According to some authors (Damgaard et al., 2009), in terms of environmental management, metal recycling processes involves reduced emissions of greenhouse gases – thus significantly contribute to the environmental protection.

Electrochemistry studies how electricity causes chemical changes, while chemical changes are reflected in electricity production. This interaction creates a huge variety of processes, from heavy industry to the batteries industry and to biological phenomena (Bockris, 1972; Walsh & Reade, 1994).

One of the main directions of electrochemistry is in the environmental protection area and addresses to: monitoring various substances that pollute or affect the environment, removal of pollutants of any kind, production and transport of energy, etc. (Iorga et al., 2009).

The main purpose of treating wastewater containing metal ions is the removal of these ions in order to reach the concentration required by law for those waters to be discharged. Process cost can be reduced if the metal can be recovered (especially for valuable metals) (Mirica et al., 2006; Rajeshwar & Ibanez, 1997).

3. Overview of methods for recovery of metal ions

Removal of metal ions from aqueous solutions may be made by conventional or electrochemical methods. In the treatment of effluents containing metal ions there are more classical approaches that can be used individually or combined (Rojanschi et al., 1997; Walsh & Reade, 1994). These are:

- diluting of the solutions to the allowed legal limit;
- mixing solution until a neutral pH;
- precipitation with a cheap precipitating agent;
- ion exchange;
- membrane separation: reverse osmosis, ultrafiltration, electrodialysis;
- storage and transport to specialized companies;
- selective solvent extraction (especially with macrocyclic compounds like crown ethers and functionalized calixarenes).

Electrochemical methods used to treat solutions with metal ions can be recuperative or destructive. Destructive electrochemical methods can be applied both to the inorganic and organic pollutants, which can be subject to reduction or oxidation. Processes are generally conducted on inert electrodes and involve complex studies because, apart from the dominant process from the electrode, parallel reactions occur due to the by-products.

Electrochemical methods for the recovery of metal ions from dilute solutions are receiving an especially attention in recent years due to difficulties encountered in treatment of contaminated waters, which often can't be destroyed or removed efficiently by conventional methods. They refer to the treatment or pre-treatment of water containing ions of heavy metals. Electrochemical methods for their recovery are based on the discharge of the metal ions at the cathode, when the reaction conditions are ensured by an appropriate design of the electrochemical reactor (Iorga et al., 2006).

Electrodeposition is a process of separation and deposition of an element or compound on an electrode, by passing an electric current between two electrodes that are immersed in an electrolyte (electrolysis). The deposition of ions can be selectively made, depending on the current or voltage applied between the electrodes. Separation and identification of metal ions from a solution can be achieved as a function of normal potentials.

Electrochemical recovery of copper from the industrial effluents is necessary for two reasons, namely:

- the improvement of the environmental conditions, in order to accomplish the environmental requirements and integration into European technology standards regarding the content of metal ions allowed in the industrial effluents;
- increasing the use of raw materials by the revaluation of the recovered metals, according to the decline of the global reserves.

4. Methods for the enhancement of mass transfer processes at the recovery of metallic ions by electrodeposition

The process of electrodeposition of metals is extremely complex, because in addition to the electrochemical steps, adsorption and mass transport, also formation and dissociation equilibria of complex, incorporation of impurities and/or additives intentionally introduced into the bath are involved.

Experimental investigation of the kinetics of cathodic deposition of the metals is a complicated operation due to characteristic features of this process. During electrolysis the cathode surface does not remain intact, it is constantly changing while the metal deposit occurs. Growth essentially depends both on the nature of the metal deposit and the conditions of electrolysis (Antropov, 2001).

Four main stages are encountered at the discharge of a metal ion (Facsko, 1969):

- cation transport by diffusion, migration or convection from the solution on the surface of the electrode – mass transport;
- desolvation of the solvated cation or dissolution of the complex ion into a simple ion – the chemical reaction;
- neutralization of the cation on the cathode surface – charge transfer;
- transition of the atom in the steady form of deposited matter (the bonding of hydrogen atoms in molecules or the input of the metal atoms into the crystalline lattice) – crystallization.

The above mentioned stages are presented in Figure 1:

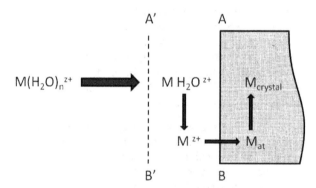

Fig. 1. Scheme of the deposition of the metal ions on the surface of the cathode: AB – surface of the electrode; A'B' – reaction zone border.

where:
AB represents the surface of the electrode;
A'B' – layer in the near vicinity of the electrode (also named reaction zone);
$M(H_2O)_n^{z+}$ – the solvated ion;
M^{z+} – simple ion.

The vast majority of the electrodeposition processes is carried out using a current source to obtain a constant current to the cathode. Optimum current density range for which satisfactory deposits are obtained is much lower than the limiting current density (Pletcher & Walsh, 2001).

Diffusion layer thickness, δ, is defined by the Nernst diffusion layer model, as shown in Figure 2. This model assumes that the concentration of the M^{z+} ions in the solution is c_a until the distance δ to the electrode surface and then decreases linearly to c_s, at the electrode surface. In this model it is assumed that the liquid layer thickness δ is practically stationary. At a distance from the surface greater than δ, reactant concentration is assumed to be equal to the one from the mass of the solution. To reach the electrode surface, M^{z+} ions must pass through the diffusion layer. At these distances, agitation of the solution becomes effective.

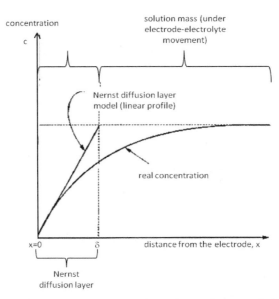

Fig. 2. Concentration profile near cathode describing the mechanism of electrodeposition process under electrode-electrolyte movement (Schlesinger & Paunovic, 2000).

When the limiting current density values reach the maximum, M^{z+} species are reduced as they reach the electrode surface. Under these conditions, the concentration of the reactant M^{z+} at the electrode is zero, and the deposition rate is controlled by the transport rate of the M^{z+} reactant to the electrode.

For this reason, the rate determining step of the process is the transport of M^{z+} ions from solution to the cathode surface. The development rate of the transport process depends on the Nernst diffusion layer (δ), which is included in the mass transport coefficient (k_m), the two quantities depend at their turn on the speed of stirring of the solution near the electrode. Thus, the stirring speed controls the diffusion limiting current density (i_L), i.e. the current density concentration where the concentration of metal ions on the cathode surface tends to zero ($c_{1s} \to 0$). The above mentioned quantities are related by equations (1) and (2):

$$i = -zFD\frac{c_{1a} - c_{1s}}{\delta} \tag{1}$$

$$i_L = -zFD\frac{c_{1a}}{\delta} = zFk_m c_{1a} \tag{2}$$

where:
i represents current density;
i_L – limiting current density;
D – diffusion coefficient of metallic ions;
k_m – mass transport coefficient;
δ – Nernst diffusion layer thickness;

c_a – solution concentration;

c_s – concentration at the surface of the electrode.

Transport overvoltage usually occurs only at low concentrations and/or high current densities, as given by equation (3):

$$\eta_t = \frac{RT}{zF} \ln\left(1 - \frac{i}{i_L}\right) \tag{3}$$

At the current crossing and the deposition of the cations, the space in the immediate vicinity of the cathode becomes poorer in ions and a concentration gradient occurs. Consequently, the cations diffusion within the solution to the cathode surface begins (Antropov, 2001).

For a given current density and concentration, transport overvoltage is lower even the higher value of the $K = \dfrac{zFD}{\delta}$ constant.

Value of the K constant can be increased either by raising the diffusion coefficient D, or by decreasing the diffusion layer thickness, δ. Increasing the diffusion coefficient (D) can be done by heating the solution. Decrease in diffusion layer thickness (δ) is achieved by solution agitation, as shown in Figure 2. So, for a current density and a given concentration, both heating and stirring of the solution lead to a lower value of the transport overvoltage.

Thus, for a given electrolyte, the deposit increasing is favoured by increasing the current density by:

- high concentrations of the dissolved metal;
- high temperature;
- relative movement between the electrode and electrolyte.

The first two options are limited in practice by the following reasons:

- metal concentration is restricted by the solubility, the cost of metal (for precious metals) and considerations of storage (and/or disposal) and effluent treatment;
- too high temperatures can aggravate corrosion problems of the equipment used in the process, losses by evaporation, chemical decomposition (e.g. when using chemical additives), prolonged duration of cooling, not to mention high energy costs for heating.

In electrochemistry, one of the main ways to increase the mass transfer of various processes is the relative movement of the electrode-electrolyte. Besides allowing the use of higher current densities and thereby increasing the production rate and improving the flow regime, it can also help for the removal of the air or of the generated hydrogen gas, or to provide a steady pH and temperature in the cathode area.

According to Gomaa et al., using the ultrasounds and vibrations to improve the mass transfer in electrochemical processes has attracted the interest mainly due to the ability of reducing the diffusion controlled processes which limit the maximum allowable current density and the energetic efficiency of the cell. This case is specific to dilute solutions where the low concentration of the active species would require supplementary actions, like the use of electrodes with higher surface or the recirculation of the electrolyte solution. For electrodeposition, metal recovery from depleted solutions and wastewater treatment, the controlled vibration technique has been applied (Gomaa et al., 2004).

The main processes applied for the improvement of electrode-electrolyte relative movement are: the use of turbulence promoters, particle fluidized bed, electrolyte recirculation, mechanical stirring of the electrolyte, electrode spinning, vibration of electrode/electrolyte, ultrasonication of electrode/electrolyte.

4.1 The use of turbulence promoters

A method for the increasing of the mass transfer at electrolytic processes that occur with release of bubbles is the induction of turbulence near the surface electrode (Vilar et al., 2011).

This alternative for the improvement of the mass transfer was applied to a system in which the electrochemical reaction on the surface of the electrode is controlled by diffusion and by the release of gas bubbles (Vilar et al., 2011). It was concluded that the mechanism leading to an increased mass transfer varies depending on the type and orientation of the metal electrode and on the electrolyte flow direction. Thus, the mass transfer is improved when the geometry of the electrode does not inhibit the release of bubbles that are generated by the electrochemical process. It was also investigated the effect of the use of triangular/semi-circular/cylindrical promoters that are placed on the bottom of a rectangular electrolytic cell. The results show that the mass transfer coefficient increases with the increasing height/radius ratio of the promoter (Venkateswarlu et al., 2010).

4.2 Use of fluidized bed particles

By using particles in fluidized bed the increasing of the surface and the intensifying of the solution movement towards the electrode are achieved. It may be achieved through a large number of metal electrodes or electrode in filling form (from small pieces) in fixed or fluidized bed. Another option is to deposit the metal on a cathode-shaped grille, which during electrolysis is surrounded by a fluidized bed consisting of small glass beads, their movement continuously destroying the polarization film from the electrodes surface.

4.3 Electrolyte recirculation

Recirculation processes of the electrolyte by pumping determine the increase of flow rate or the degree of mixing of the electrolyte. Electrolysis cells do not contain electrodes or other moving parts, which determine reduced power consumption and mechanical wear. Dynamic flow regime allows the use of relatively high current densities even for small concentrations and also the achievement of an advanced degree of exhaustion of the solution subjected to electrolysis (Pletcher & Walsh, 2000). Cathodes on which the recovered metal is deposited are periodically removed from the electrolytic bath and replaced with new plates.

4.4 Mechanical stirring of the electrolyte

Mechanical mixing processes of the electrolyte are characterized by a strong movement of the solution that is subjected to electrolysis, which is achieved by using mechanical stirrers. The cell has an optimum design for a turbulent flow of the solution. The stirrer consumes less energy than a circulator pump, which would achieve the same degree of agitation.

There are used flat cathodes, with simple forms. To avoid the complete coverage of cathodes, they are provided with plastic masks, placed on the edges.

4.5 Spin of electrodes

Rotating process of the electrodes is a way to achieve the relative movement of the electrolyte to the electrodes by rotating one of them, usually the anode. There are mentioned experiments (Eisenberg et al., 1954) that studied the effect of the rotational speed on the rate of mass transfer in the case of rotating electrode. It is also mentioned (Eisenberg et al., 1954) the use of rotating electrodes in order to study the speed of dissolution of Mg and its alloys in HCl solutions. Thus, they determined that dissolution rates increase with rotation speed (when using a diluted solution of HCl). If more concentrated solutions of acids are used, reaction rates are completely independent of the speed of rotation, as the shaking effect produced by the release of H_2 bubbles to surface prevails. In some cases, a rotating disc-shaped electrode is combined with high-speed radial flow of electrolyte.

Recendiz et al. had used rotating cylinder electrodes for the study of mass transport in a turbulent flow regime. Mass transport control is imposed by the speed of rotation of the inner cylinder and by the applied limiting current density. Studies regarding the mass transport at RCE (for the case of copper electrodeposition) were performed in turbulent regime (Re > 100). Experimental determination of mass transport coefficients was made by using electrolysis experiments at a given potential (Recendiz et al., 2011).

4.6 Vibrating of the electrode/electrolyte

One of the most effective methods for the enhancing of the mass transport in electrolytic cells is represented by the vibration. The vibration technique is a method of increasing the electrode-electrolyte relative movement and is defined as initiating and sustaining of a regular movement with well defined mechanical parameters (amplitude A, frequency ω) between the active surface electrode and electrolyte. The strong effect of stirring the solution layer in the immediate vicinity of the electrode determines the reducing of the diffusion layer and the limiting current density may be increased for 10-15 times. As such, current yields are significantly improved, the electrolysis voltage decreases, the specific consumption of electricity is accordingly reduced, and the electrolysis cells productivity significantly increase.

Generation of the pulsing field at the electrode-electrolyte interface can be achieved either by vibrating the electrode surface, either the volume of the electrolyte. As regards the relative motion, both methods work in the same way, but the specific energy consumption clearly favours the first procedure. This is due to the fact that, the afferent energy of the vibration manifests in the boundary layer adjacent to the electrode surface and not in the volume of solution, such as the second alternative does (Gomaa et al., 2004).

Vibrating electrode provides particularly advantageous conditions for conducting electrochemical processes controlled by mass transfer, especially due to the fact that it provides a substantial increase of the ionic species transport to and from the electrode surface (mass transfer coefficient k increases when d decreases, which happens when forced convection conditions by lowering h_c and i_{lim}).

Vibration technique can be successfully applied for dilute solutions. Vibrating electrodes can be done by using two main modes of vibration:

- longitudinal (parallel movement to the active surface);
- transverse (movement occurs perpendicular to the active surface).

Vibration of the electrode substantially modifies the boundary layer that appears at the electrode-electrolyte interface in the case of free convection, the layer that also contains the reaction product. Previous studies regarding the effect of vibrations on the mass transfer rate proved that the vibrations are an efficient method for the improvement of the mass transfer (Fakeeha et al., 1995).

Literature data present a number of researches regarding the improvement of the mass transfer by vibrations. There are many relations regarding the mass transfer and a number of applications on electrodes of various shapes. For the case of spherical vibrated electrodes, the obtained relation is equation (4) (Venkata Rao et al., 2010):

$$Sh_v = 2 + 0.24 \left(\frac{f\rho dH}{\mu} \right)^{\frac{1}{2}} \tag{4}$$

In the case of mass transfer at the copper spherical vibrated electrodes, the equation (5) was proposed (Venkata Rao et al., 2010):

$$\frac{Sh_v}{Sc^{\frac{1}{3}}} = 0.97 Re_v^{0.4} \tag{5}$$

For the case of copper and stainless steel cylindrical electrodes, the mass transfer is characterized by the relation (6) (Venkata Rao et al., 2010):

$$J_{dv} = 0.41 Re_v^{-0.38} \tag{6}$$

The influence of the sinusoidal vibrations on the mass transfer by forced convection between a solid sphere and a liquid is given by the empirical relation (7) (Venkata Rao et al., 2010):

$$\frac{Sh_v}{Sc^{\frac{1}{3}}} = 0.477 Re_v^{0.538} \left[1 + 1.05 (V - 0.06)^{1.26} \right], \tag{7}$$

where: Sh, Sc, Re are the dimensionless numbers of Sherwood, Schmidt and Reynolds criteria.

Venkata Rao et al. (2010) investigated the efficiency of the mass transfer in the case that a disc electrode was vibrated. Vibrating the electrode leads to instability into the electrolyte (around the electrode surface) due to the wave's formation and results in a reduced thickness of deposited film on the electrode surface. Thus, the vibration rate of the electrode also influences the thickness of the deposited film and the transfer coefficient. The highest the vibration rate, the highest the mass transfer coefficient is.

Takahashi and collaborators (Takahashi et al., 1992) have studied the effect of the vibrations on local and global mass transfer for the cylindrical tube vibration in a stationary electrolyte.

The vibration effects were mainly observed on the local transfer coefficients, in the side positions of the electrode. Noordsiji and Rotte (Noordsiji & Rotte, 1961) studied the vibration effect on the mass transfer for a spherical nickel electrode and have presented the relationships that describe the mass transfer criteria.

Takahashi and co-workers (Takahashi & Endoh, 1989) have studied the effect of the vibrations on spherical-, disc- or cylinder-shaped electrodes. Noordsiji and Rotte (Noordsiji & Rotte, 1961) have studied the mass transfer at vertical cylindrical vibrated electrodes at large amplitudes. The same study, but for horizontally electrodes can be found in the study of Venkateswarlu and co-workers (Venkateswarlu et al., 2002).

Other authors (Rama Raju et al., 1969) have studied the ionic mass transfer in mixed convection conditions, for the specific case of pulsating flow. In terms of energy consumption compared to vibration mechanical parameters (A, ω), it was found (Takahashi et al., 1993) that an increase in amplitude and frequency has similar effects on the mass transfer enhancement.

Gomaa et al. investigated the longitudinal vibration effect on the improvement of the mass transfer. It may be said that, the mass transfer rate at vertical surfaces can be significantly improved by vibrating in a longitudinal direction parallel to their surfaces. The degree of the improvement of mass transfer coefficient depends on the amplitude and frequency of the vibrations and of the length of the electrode. An improvement up to 23 times of the mass transfer was recorded. Average coefficient of mass transfer at vibrating electrodes can be predicted by using a conventional mass transfer equation for plane surfaces in that a pseudo-equilibrium rate is used (Gomaa et al., 2004).

The improvement of the mass transfer by vibrations that are applied to electrochemical process leads to a significant increase of the performance of the process. This reduces the polarization of the concentration, together with the acceleration of the deposited metal, the degree of dissolution and also with increased current yields. It facilitates the formation of fine metal deposits (as powders) and release of gas bubbles. It is also noted a significant reduction in the volume of processing units necessary for the electro-processing, which determines lower processing costs.

Fakeeha et al. have studied the effect of oscillatory movement on the mass transfer at vibrating electrodes, causing the current limit values for each applied frequency of the frequency range (0÷26 Hz). Around the value of 13 Hz, a sudden drop in the limit current values occurs, and then its growth is resumed.

The effect of frequency on the mass transfer rate can be explained by Stoke solution for the case of the flow near a vibrating electrode (Fakeeha et al., 1995):

$$U(\eta, t) = A e^{-\eta} \cos(Ft - \eta) \tag{8}$$

$$\eta = \frac{yF}{2v} \tag{9}$$

where:

U is the electrolyte flow velocity induced by vibration;

y – distance to the electrode surface;

t – time;

F – frequency of the vibration;

A – amplitude of the vibration;

v – cinematic viscosity.

The increase of the mass transfer seems to be due to the decreased thickness of the adjacent layer to the electrode surface which is induced by vibration. The vibration-induced flow seems to be stronger than the continuous flow of electrolyte, thereby improving the mass transfer coefficient. Although the influence of hydrodynamics on the transport mechanisms from the electrochemical cells has been previously investigated, the situation is different for cells in which vibrations are used to improve mass transfer; there are still some deficiencies in understanding the various mechanisms that are contributing to the overall improvement of the mass transfer coefficient.

4.7 Ultrasonication of the electrode/electrolyte

Influence of an ultrasonic field on electrochemical processes is manifested as follows: in a fluid, the propagation of an ultrasonic pressure wave produces the cavitation effect, consisting of the formation of bubbles containing gas and liquid vapours. These bubbles, at low intensities of the ultrasonic field, oscillate in the electrolyte volume and they essentially contribute to the mass transfer and to the changes of the diffusion layer around the electrodes.

The use of ultrasonic techniques in electrochemistry has several advantages, such as a limited accumulation of gas bubbles at electrode and a high efficiency of the ions transport through the double layer. In addition, there is a continuous cleaning and activation of the electrode surface (http://www.coventry.ac.uk/researchnet/Sonochemistry/Pages/Sono-chemistry.aspx).

All these improvements have beneficial effects on the diffusion processes, of the yields and the rate of electrodeposition processes and they are due to the cavitation effect applied to the electrolyte (Mallik & Ray, 2009). The electrode surface causes the bubbles collapse leading to the formation of a liquid jet that is directed toward the surface with high speed. Thus, the double layer destruction and, consequently, the improvement of the mass transfer occur.

There are two methods for the use of ultrasound in electrochemistry: ultrasonication of working electrode by using the ultrasonic transducer or ultrasonication of the electrolyte.

The literature mentioned the ultrasonic-assisted electrodeposition of various metals and alloys. This process provides a considerable improvement of the mass transfer compared to the conventional electrodeposition methods (Grunwald, 1995). It was found that ultrasonic agitation during electrodeposition produces smoother and more compact deposits, with increased corrosion resistance.

In the case of electrodepositions, the influence of intensity and frequency of ultrasound on the deposits hardness, particle size and morphology of surfaces is studied. Application of ultrasound to electrochemical processes increases the transport of the electroactive species to/from the electrode both in the diffusion layer and inside the double layer by electrolyte stirring due to cavitation and pressure waves.

At the overcoming of the field intensity above a certain threshold value, the phenomenon of collapse of these bubbles (implosion) appears, leading to local conditions of extremes temperature and pressure. The phenomenon is associated with the microflow or with the microcurrents and with the emergence of elevated electric fields. This leads to mechanical phenomena of corrosion, the fragmentation of solid particles and emulsifying of immiscible liquids. In terms of electrochemistry, the above mentioned effects are studied in terms of their influence on electroplating, electrodeposition and electrosyntheses, increased reaction rates and change of the electrodeposition quality (Doche et al., 2001; Hardcastle et al., 2000; Mason and Lorimer, 2002; Walton and Phull, 1996).

For the electrodeposition of copper and zinc, the fewer gaps in the structure of the metal is due to crushed crystals or pushed into holes by air pressure or by shock wave pressure. Copper electrodeposition has been performed in the narrow channels of the electronics and electrical parts. The phenomenon is attributed to cavitation and to the improvement of mass and heat transfer and copper electrodeposition, while hydrogen is discharged.

Coverage of aluminium includes the application of ultrasound in the electrolysis baths that contain molten salt, in order to remove oxygen. In the case of nickel electrodeposition, ultrasonication influences the crystallization process through the effect of cavitation in the electrolyte, the quality of the deposit being also greatly influenced from the current density.

Using the ultrasonic stirring together with plating, much durable black platinum deposits than platinum wires without stirring are obtained. This provides considerably larger surfaces of the precious metal wire, by increasing the flow of $PtCl_6^-$ ions, due to ultrasonic agitation. In the case of silver electrodeposition, due to the effect of ultrasonic cavitation, the adherence is improved and deposits cracks are removed. It results a mate aspect of the silver surface. In the case of chrome plating, the ultrasonic bath and the appearance of cavitation lead to the obtaining of deposits with higher hardness, but with several cracks. In contrast, the obtained deposits are more shiny and spotless.

For the electrodeposition of iron, the deposits obtained by ultrasonication have higher hardness, reduced porosity and surface staining. The ultrasound-assisted electrodeposition of Ni-Zn alloys favours the accumulation of more shiny zinc deposits.

Application of ultrasound for the electrodeposition of Ni-Fe alloys leads to an increased iron content, lack of roughness and increased hardness of the deposit. Electroplating with noble metals like Au, Rh, Ru and corresponding alloys from electrolyte solution containing organic compounds in order to prevent the corrosion of the substrate is mentioned.

By using ultrasound in electrochemical processes electrodynamics of the electrolysis cell is improved. The effects are similar to those produced by known hydrodynamic techniques, such as agitation, recirculation or vibrating. There can't be neglected the thermal effect of ultrasonication that leads to the warming of the solution. Unlike other methods for the increasing of the electrode processes, in this case the cavitation effect is involved.

5. Comparative analysis of procedures to enhance the relative electrode-electrolyte movement to the electrochemical deposition of metal

Laboratory studies regarding copper electrodeposition processes are presented below. Based on previous works (Iorga et al., 2011), various possibilities have been investigated to improve the mass transfer in the electrodeposition of Cu from $CuSO_4/H_2SO_4$ solutions.

In order to observe the influence of the electrode-electrolyte relative movement, the electrodeposition processes were investigated in several operating modes, namely: mechanical stirring of the electrolyte, vibration of the electrode and electrolyte ultrasonication regime. Thus, comparisons among the obtained results of the electrodeposition under stirring, vibrating and ultrasonic conditions were made.

Experiments were performed in a standard cell with three electrodes. A number of solutions of different concentrations of $CuSO_4$ (0.05M ÷ 0.65M concentration range) in 0.25, 0.5 and 0.75 M H_2SO_4 have been used. The polarization curves were drawn with a potentiostat PGZ Voltalab 402.

A volume of 50 ml $CuSO_4/H_2SO_4$ solution, a copper working electrode with an active area of 0.05 cm², an Ag/AgCl (sat. KCl) reference electrode and an auxiliary electrode (Pt wire) with an area of 0.25 cm² were used for the experimental determinations. Scanning speed was set at 20 mV/s in a potential range from 100 mV ÷ -1100 mV. To ensure the reproducible results and to ensure constant conditions, the working temperature was chosen the ambient temperature, 22÷24 °C.

During the experiments, several parameters (ultrasonic bath functional parameters, frequency of vibration of the electrode) have been varied and the optimal values for each operating mode were determined.

For the deposition of copper under mechanical stirring a magnetic stirrer was used.

To achieve the vibrating regime, a vibration system for static elements / electrodes patented in INCEMC Timisoara (Buzatu et al., 2011) was used. The system covers a wide range of frequencies (with a very good resolution, at least 1 Hz) and has the possibility of using a wide range of waveforms (sine wave, rectangular, triangular or saw wave, etc.).

Ultrasonication regime used an ultrasonic bath from FALC Instruments, Italy, at 25 °C and 59 kHz.

5.1 Results of the mass transfer improvement methods that were applied to the electrolyte solution

5.1.1 Ultrasonication of the electrolyte

In order to determine the optimum conditions for copper ultrasound electrodeposition, variations of the functional parameters of the ultrasonication bath have been performed.

Polarization curves of copper electrodeposition process from a 0.05 M $CuSO_4$ solution have been recorded for the both working frequencies of the bath, namely 40 and 59 kHz, and are presented in figure 3.

As one can observe from figure 3 the difference between the results obtained at two different values of the working frequencies is insignificant. For further determination was chosen a working frequency of 59 kHz (due to the slight pronounced deposition plateau).

Another parameter that can be varied is the power at which the ultrasonication bath works. The results are presented in figure 4:

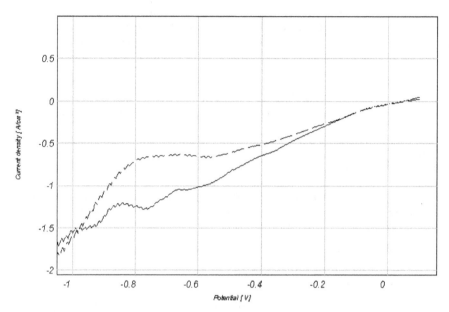

Fig. 3. Polarization curves at copper electrodeposition from a 0.05 M CuSO$_4$ solution, for different working frequencies: _____ 40 kHz; _ _ _ 59 kHz.

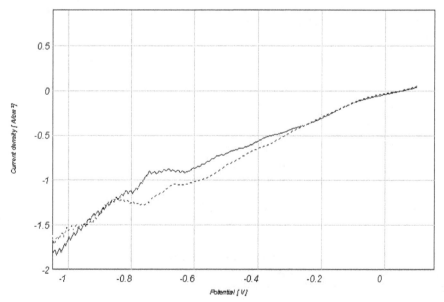

Fig. 4. Polarization curves at copper electrodeposition from a 0.05 M CuSO$_4$ solution, for different percents of bath power: _____ 50 %; .... 75 % (from 285 W, total bath power).

The results presented in Figure 4 shows that is no significant difference between the two values of the power. As a consequence, for further determination was chosen the option that uses 50 % of the total power (due to the slight pronounced deposition plateau).

5.1.2 Comparison between the ultrasonication and the mechanical stirring of the electrolyte

There were two methods for the mass transfer improvement that were applied to the electrolyte solution, namely the mechanical stirring and the ultrasonication. Both assays were applied to a 0.05 M $CuSO_4$ solution and the results are presented in figure 5.

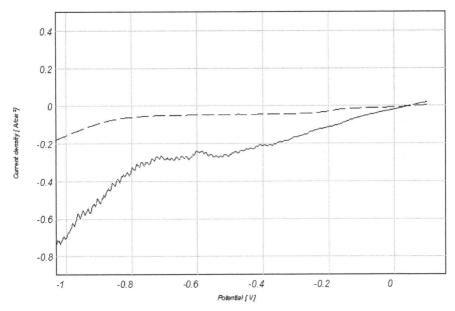

Fig. 5. Polarization curves at copper electrodeposition from a 0.05 M $CuSO_4$ solution: _____ ultrasonication of the electrolyte; _ _ _ mechanical stirring of the electrolyte.

From figure 5 it may be seen that the ultrasound-assisted electrodeposition of copper is favoured (if the current intensity is taken into account), but also one of the major drawback of the method appears, namely a less pronounced deposition plateau. This is due the fact that the ultrasounds also affect the electrode that is immersed in the electrolyte solution, and, consequently, it alters the metal deposition.

5.2 Results of the mass transfer improvement methods that were applied to the electrode

The experiments were conducted by using a vibrating electrode system that allows variation of both frequency and waveforms. The variable parameter was chosen to be the vibration frequency, while the waveform was set to the sine one. For the first set of frequencies (below 100 Hz), the results are presented both in graphic (figure 6) and table (table 1) form.

5.2.1 Frequencies < 100 Hz

The data from Figure 6 and Table 1 show that, better results are obtained for the two of lowest employed frequencies, namely 30 Hz and 20 Hz.

Frequency (Hz)	i (mA/cm²)	ε (V)
20	-72.172	-0.42689
30	-79.750	-0.44258
40	-62.471	-0.43159
60	-55.825	-0.46141
80	-39.875	-0.47397

Table 1. Limit current densities and potential values (frequencies < 100 Hz).

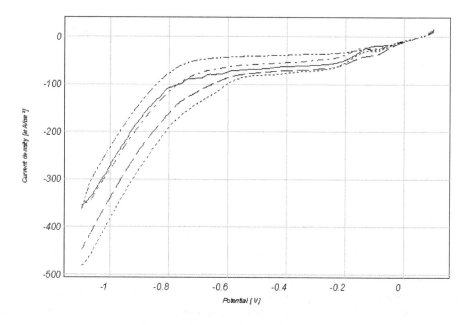

Fig. 6. Polarization curves at copper electrodeposition from a 0.05 M CuSO₄ solution, at different frequencies (<100 Hz): _ _ _20 Hz; .....30 Hz; ___40 Hz; ¯¯¯¯¯ 60 Hz; ▬ ∙▬ ∙∙ 80 Hz.

5.2.2 Frequencies > 100 Hz

Results from Figure 7 and Table 2 show a significant difference between the obtained results at a frequency of 100 Hz and all other three higher frequencies. It may be concluded, together from the data presented in Figures 6 and 7, that the frequency of 100 Hz is the most favourable for the copper electrodeposition.

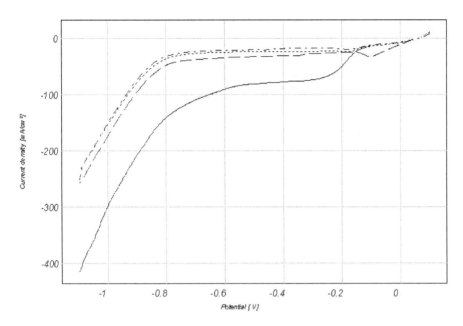

Fig. 7. Polarization curves at copper electrodeposition from a 0.05 M CuSO₄ solution, at
different frequencies (>100 Hz): ____100 Hz; _ _ _150 Hz; .....200 Hz; ‑‑‑‑‑‑ 300 Hz.

Frequency (Hz)	i (mA/cm²)	ε (V)
100	-78.772	-0.44415
150	-32.385	-0.49908
200	-24.462	-0.48339
300	-16.679	-0.41433

Table 2. Limit current densities and potential values (frequencies > 100 Hz).

5.3 Comparison between the mass transfer enhancement methods applied to the electrode and to the electrolyte, respectively

The polarisation curves obtained at the electrodeposition of copper when the three above presented working regimes are presented all together in the Figure 8 and Table 3:

The allure of the polarization curves presented in Figure 8 clearly shows an increased efficiency of the ultrasound-assisted working regime. Also, it may be observed that in the case of a more concentrated CuSO₄ solution, the plateau that corresponds to the electrodeposition of metals is better defined than in the case where electrolytes with a lower concentration in CuSO₄ were used (see Figure 5). The almost double performance of the ultrasound-assisted working regime towards the vibrating electrode assay (in terms of i values) recommends this method for further investigations as regards its use at larger scales.

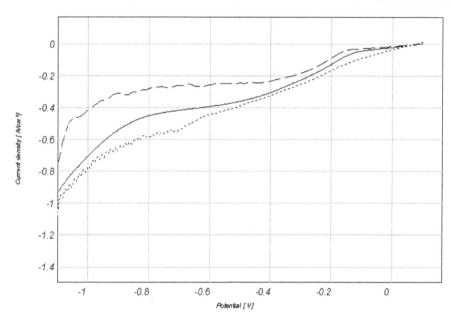

Fig. 8. Polarization curves at copper electrodeposition from a 0.35 M CuSO₄ solution: _ _ _
mechanical stirring of the electrolyte; ___ vibrating electrode; ..... ultrasound-assisted
electrodeposition of copper.

Working regime	i (A/cm²)	ε (V)
Mechanical stirring	-0.2609	-0.645
Vibrating	-0.4033	-0.693
Ultrasonication	-0.5456	-0.751

Table 3. Limit current densities and potential values.

6. Conclusions

The paper presents the main methods for the increasing of mass transport during the electrochemical deposition of the metals. The enhancement of the electrode processes represents the key problem of the hydroelectrometallurgical processes and, as in the case of diluted solutions, the processes are controlled by the transport of the reactant species to the electrode. The literature study presented underlines the main methods (and especially the ones with industrial applications) that are applied for the enhanced mass transfer: the use of turbulence promoters, particle fluidized bed, mechanical stirring of the electrolyte, electrode spinning, electrode/electrolyte vibration and electrode/electrolyte ultrasonication.

The experimental part deals with three of the most widely applied techniques for the increasing of the mass transfer. A diluted solution of CuSO₄ was chosen as electrolyte and three different regimes, namely mechanical stirring, ultrasonication and vibrating were developed. The results show a better performance of the ultrasound-assisted electrodeposition of copper, followed by the vibrating electrode method and by the mechanical stirring of the electrolyte, respectively.

7. References

Antropov, L. I. (2001). *Theoretical Electrochemistry*, University Press of the Pacific, Honolulu, Hawaii

Bockris, J.O'M. (1972). *Electrochemistry of Cleaner Environments*, Plenum Press, New York

Buzatu, D., Iorga, M., Mirica, M.C., Urmosi, Z., Pop, R., Balcu, I., Mirica, N. (2011). Sistem de vibrare a elementelor statice/electrozilor cu aplicaţii în procese chimice şi electrochimice. Patent A/00478, OSIM Bucuresti

Damgaard, A., Larsen, A.W., Christensen, T.H. (2009). Recycling of metals: accounting of greenhouse gases and global warming contributions. *Waste Management & Research*, Vol. 27, pp. 773-780

Doche, M. L., Hihn, J. Y., Touyeras, F., Lorimer, J. P., Mason, T. J., Plattes, M. (2001). Electrochemical behaviour of zinc in 20 kHz sonicated NaOH electrolytes. *Ultrasonics Sonochemistry*, Vol. 8, pp. 291-298

Eisenberg, M., Tobias, C. W., Wilke, C. R. (1954). Ionic mass transfer and concentration polarization at rotating electrodes. *Journal of the Electrochemical Society*, Vol. 101, pp. 306-320

Emery, A., Williams, K. P., Griffiths, A. J. (2002). A review of the UK metals recycling industry. *Waste Management & Research* 20: 457-467

Facsko, G. (1969). *Tehnologie electrochimică*, Editura Tehnică, Bucureşti

Fakeeha, A. H., Abdel Aleem, F. A., Inamah, I. A. (1995). Mass transfer at vibrating electrodes in continuous flowing electrolyte systems. *The 4th Suadi Engineering Conference*, Vol. V, pp. 81-88

Gomaa, H., Al Taweel, A. M., Landau, J. (2004). Mass transfer enhancement at vibrating electrodes. *Chemical Engineering Journal*, Vol. 97, pp. 141-149

Grünwald, E. (1995). *Tehnologii moderne de galvanizare în industria electronică şi electrotehnică*, Ed. Casa Cărţii de Ştiinţă, Cluj-Napoca

Hardcastle, J. L., Ball, J. C., Hong, Q., Marken, F., Compton R.G.; Bull S.D.; Davies S.G. (2000). Sonoelectrochemical and sonochemical effects of cavitation: correlation with interfacial cavitation induced by 20 kHz ultrasound. *Ultrasonics Sonochemistry* 7: 7-14

Iorga, M., Mirica, N., Mirica, M.C., Dragoş, A. (2006). Cathodic processes improvement in silver ions recovery from waste photografic solutions, In: *Proceedings of Romanian Chemistry Conference*, Calimanesti-Caciulata, Valcea, Romania, P.S.IV-18, pp.342

Iorga, M., Vaszilcsin, N., Balcu, I., Mirica, M.C., Dragoş, A. (2007). Silver ions recovery by electrodeposition, In: *Proceedings of INCEMC's Anniversary Simposium*, Timişoara, Romania, S2.P-4, pp.35

Iorga, M., Mirica, M.C., Buzatu, D. (2009). Monitoring autonomous station with applications in environment protection and photovoltaic systems installation, In: *Proceedings of Main Group Chemistry Conference*, – ZING, Cancun, Mexico, P10, pp.45

Iorga, M., Mirica, M.C., Balcu, I., Mirica, N., Rosu, D. (2011), Influence of Relative Electrode-Electrolyte Movement over Productivity for Silver Recovery from Diluted Solutions. *Journal of Chemistry and Chemical Engineering* – JCCE, Vol.5, 4, pp.296-304

Kuhn, A. T. 1971. *Industrial Electrochemical Processes*, Elsevier, Amsterdam

Mallik, A., Ray, B. C. (2009). Morphological study of electrodeposited copper under the influence of ultrasound and low temperature. *Thin Solid Films*, Vol. 517, pp. 6612-6616

Mason, T. J., Lorimer, J. P. (2002). *Applied Sonochemistry. The Uses of Power Ultrasound in Chemistry and Processing*, Wiley VCH, Weinheim

Mirica, N., Dragoş, A., Mirica, M. C., Iorga, M., Macarie, C. (2006). *Metode electrochimice de recuperare a ionilor metalici din soluţii*, Ed.Mirton, Timişoara

Noordsiji, P., Rotte, J. W. (1961). Mass transfer at and to a vibrating sphere. *Chem. Eng. Sci.*, Vol. 22, pp. 1475:1481

Pletcher, D., Walsh, F. C. (2000). *Industrial Electrochemistry*, 2nd Edition, Springer

Rajeshwar, K., Ibanez, J.G. (1997). *Environmental Electrochemistry. Fundamentals and Application in Pollution Abatement*, Academic Press, Inc., San Diego

Rama Raju, C. V., Sastry, A. R., Raju, G.J. V. J. (1969). Ionic mass transfer at vibrating plates, *Indian J. Technol.*, Vol. 7, pp. 35–37

Recendiz, A., Leon, S., Nava, J. L., Rivera, F. F. (2011). Mass transport studies at rotating cylinder electrode during zinc removal from dilute solutions. *Electrochimica Acta*, Vol. 56, pp. 1455-1459

Rojanschi, V., Bran, F., Diaconu, G. (1997). *Protecţia şi ingineria mediului*, Ed. Economică, Bucureşti

Schlesinger, M., Paunovic, M. (2000). *Modern Electroplating*, John Wiley & Sons Inc., New York

Takahashi, K., Endoh, K. (1989). Effect of vibration on forced convection mass transfer. *J. Chem. Eng. Jpn.*, Vol. 22, pp. 120-124

Takahashi, K., Hirano, H., Endoh, K., Imai, H. (1992). Local mass transfer from a single cylinder and tube banks vibrating sinusoidally in a fluid at rest. *J. Chem. Eng. Jpn.*, Vol. 25, pp. 678-683

Takahashi, K., Mori, S. K. S., Tanimoto, A. (1993). Some features of combined convection mass transfer for pulsating fluid flow in a pipe. *Kagaku Kogaku Rombunshu*, Vol. 19, pp. 127-130

Venkata Rao, P., Venkateswarlu, P. (2010). Effect of vibrating disc on ionic mass transfer in an electrolytic cell. *International Communications in Heat and Mass Transfer*, Vol. 37, pp. 1261-1265

Venkateswarlu, P., Chittibabu, N., Subba Rao, D. (2010). Effect of turbulence promoters on ionic mass transfer in a rectangular electrolytic cell. *Indian Journal of Chemical Technology*, Vol. 17, pp. 260-266

Venkateswarlu, P., Jaya Raj, N., Subba Rao, D., Subbaiah, T. (2001). Mass transfer conditions on a perforated electrode support vibrating in an electrolytic cell. *Chemical Engineering and Processing: Process Intensification*, Vol. 41, pp. 349-356

Vilar, E. O., Cavalcanti, E. B., Albuquerque, I. L. T. (2011). A Mass Transfer Study with Electrolytic Gas Production, *Advanced Topics in Mass Transfer*, Mohamed El-Amin (Ed.), ISBN: 978-953-307-333-0, InTech

Walsh, F. C., Reade, G. W. (1994). Electrochemical techniques for the treatment of dilute metal-ion solutions, in *Studies in Environmental Science 59. Environmental Oriented Electrochemistry*, Edited by C.A.C. Sequeira, Elsevier, Amsterdam

Walton, D. J., Phull, S. S. (1996). Sonoelectrochemistry, in *Advances in Sonochemistry*, JAI Press, London

Zhang, S., Forssberg, E. (1998). Mechanical recycling of electronics scrap – the current status and prospects. *Waste Management & Research*, Vol. 16, pp. 119-128

3

Oxidative Hydrometallurgy of Sulphide Minerals

F. R. Carrillo-Pedroza, M. J. Soria-Aguilar, E. Salinas-Rodríguez,
A. Martínez-Luevanos, T. E. Pecina-Treviño and A. Dávalos-Sánchez
Autonomous University of Coahuila,
Mexico

1. Introduction

Sulphide minerals are one of the most important sources of value metals, such as gold, silver, copper, zinc, etc. Due to the strong sulphur binding to these minerals, metals are usually extracted by pyrometallurgical route or hydrometallurgy with chemical oxidation. Of these, hydrometallurgy apparently has a lower environmental impact, which has received increased attention in last decades. The main stages of the hydrometallurgical route comprise leaching, extraction and precipitation or electrowinning. For several decades, a number of processes have been developed to leach sulphide ores and concentrates and the conditions are well established. However, there is a renewed interest in hydrometallurgical processes for copper production due to environmental issues and the increasing need to exploit mixed and low grade ores and relatively small isolated deposits.

Processing of these ores and deposits is very slow and requires a significant amount of reagents. Therefore, to make the process profitable, the treatment of large quantities of ore is required. Aqueous oxidation can be conducted under elevated temperature and pressure, but also at ambient conditions, which makes it environmentally and economically attractive. For this reason, studies to optimize aqueous oxidation and to explore more efficient oxidants have been made. However, in mining industry (especially in precious metals extraction), the use of advanced oxidation process or ozone as an oxidant has not been discussed in detail, although lab-scale experiments indicate that ozone may be an alternative to overcome economic and ecological disadvantages of aqueous extraction existing process.

In this Chapter, we will treated the use of ozone and advanced oxidation process, including microwave system, as methods to improve or to help the leaching of different sulphide minerals. For example, it is well known that ozone is a powerful oxidizing with high oxidation potential (2.07 V) compared with hydrogen peroxide (1.77 V) and chlorine (1.4 V), making it advantageous to use in several applications. Importantly, ozone can create favorable conditions to oxidize sulphide minerals in aqueous media. In this context, oxidative leaching with ozone is relevant in copper-iron sulphide and gold- and silver-containing sulphides. Moreover, oxidative leaching of coal-containing iron sulphide might also have a positive impact on coal cleaning prior to its use in energy related applications.

The hydrometallurgy of different sulphide minerals will be treated. We will discuss and analyze the lab result that we obtained with these type of minerals. Cyanidation of gold-silver pyritic minerals with ozone pre-treatment, chalcopyrite and sphalerite leaching with oxidation and microwave as complementary methods, and pyrite dissolution present in coal by oxidants aqueous media, will be treated here. In each case including aspects as chemical reactions, thermodynamics (Pourbaix´s Diagrams), kinetics and analysis of factors with statistical tools are discussed.

Statistical tool, as Factorial and Taguchi experiment´s design and analysis of variance (ANOVA) will receive a particular attention. These methods are now widely used to provide the optimal selection of parametric values based on their intraparametric interactions to accomplish a process and determine the optimum leaching conditions.

2. Fundamentals

Valuable metals are recovering worldwide relevance due to the development of a whole new range of potential applications in electronics, environmental catalysis, material science, biomedicine, among other fields with significant impact in daily life activities.

Sulphide minerals, as pyrite, FeS2, chalcopyrite and CuFeS2, are one of the most important sources of value metals, such as gold, silver, copper, zinc, etc. Due to the strong sulfur binding to these minerals, metals are usually extracted by metallurgical process of chemical oxidation.

In Extractive Metallurgy, process can be divided in Pyrometallurgy and Hydrometallurgy. Particularly, chemical oxidation can be classified generally as roasting and aqueous dissolution. Roasting under oxidizing condition is a very extensive and well established commercial technology. However, roasting has been considered as a high energy consumer technology, with stringent environmental controls on the emission of gases. Hence, aqueous chemical oxidation methods have attracted increasing attention. The aqueous oxidation can be operated under elevated temperatures and pressures or ambient conditions. Definitely, low pressure and temperature are seen as environmentally and economically attractive (Deng, 1992). Aqueous oxidation can be conducted under elevated temperature and pressure, but also at ambient conditions, which makes it environmentally and economically attractive (Deng, 1992). For this reason, studies to optimize aqueous oxidation and to explore more efficient oxidants have been made. However, in mining industry (especially in precious metals extraction), the use of ozone as an oxidant has not been discussed in detail, although lab-scale experiments indicate that ozone may be an alternative to overcome economic and ecological disadvantages of aqueous extraction existing process.

2.1 Thermodynamics of oxidation process

Ozone has a very high oxidation potential (2.07 V) compared with hydrogen peroxide (1.77 V) and chlorine (1.4 V), making it advantageous to use in several applications (Rice, 1997). Importantly, ozone can create favorable conditions to oxidize sulphide minerals in aqueous media. According to the Pourbaix or Eh – pH diagrams shown in Figures 1 and 2, the sulphide species such as pyrite or pyrrhotite (Fig. 1) and chalcopyrite (Fig. 2), can be oxidized to sulfate in presence of an oxidant such as ozone, in a pH range from 2 to 14; the oxidized products could be solids or solutions. At very acid conditions (i.e, pH < 2), it is possible to dissolve metals as Fe and Cu ions. In this context, oxidative leaching with ozone is relevant in copper-iron sulphide and gold- and silver- containing sulphides. Moreover, oxidative leaching of coal-containing iron sulphide might also have a positive impact on coal cleaning prior to its use in energy related applications. In this paper, we show the

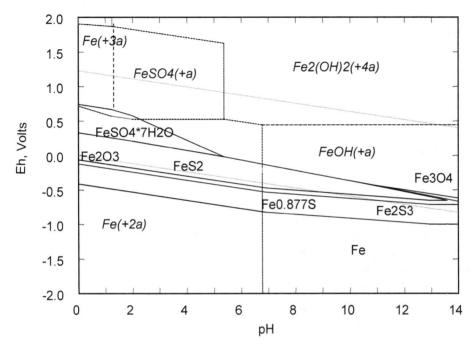

Fig. 1. Pourbaix diagram for S-Fe system, at 25 °C, (Fe) = 1 M; (S) = 1M.

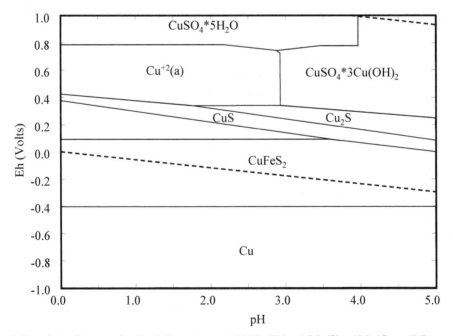

Fig. 2. Pourbaix diagram for Cu-S-Fe system, at 25 °C, (Fe) = 1 M; (S) = 1M; (Cu = 1M).

beneficial effect of using ozone on process of environmental and commercial importance, and outline the role of ozone layer in process optimization. The practical significance of the study cases is briefly discussed next.

2.2 Chemical reactions

In this context, it should emphasize the process of oxidation of sulfides as exemplified by oxidation of pyrite, one of the most abundant minerals on earth. In general, under oxidant condition and low pH, pyrite oxidation proceeds through two basic steps: In the first step, the dissolution of pyrite to ferrous ions in an acid medium proceeds through the formation of an iron-deficient or a sulfur-rich layer rather than elemental sulfur.

In the second step, further oxidation of this layer occurs, forming sulfides of lower iron content, and eventually are converted to elemental sulfur. In severely oxidizing conditions, the elemental sulfur could be oxidized to oxy-sulfuric species. Anodic reactions, such as pyrite and sulfur oxidations, are sustained by cathodic processes, which could involve oxygen, hydrogen peroxide, or even ozone reduction. The importance of this analysis is based on the fact that, under certain conditions, such as pH, redox potential, temperature, etc., the product layer is protective, thus limiting pyrite oxidation.

Despite the existing discrepancies about the exact composition of the oxidation products, the most well-known general mechanism of pyrite oxidation is described in Eq 1.

$$FeS_2 = Fe^{2+} + 2S^\circ + 2e- \tag{1}$$

Elemental sulfur is stable at low pH and redox potential and could be oxidized to sulfate by molecular oxygen and ferric ions at higher potentials (Eq 2).

$$FeS_2 + 8H_2O = Fe^{3+} + 2SO_4^{2-} + 16H^+ + 15e- \tag{2}$$

The pyrite dissolution has been characterized in the following media:

i.   in the presence of oxygen at high pressure and temperature

$$2FeS_2 + 7O_2 + 2H_2O = 2FeSO_4 + 2H_2SO_4 \tag{3}$$

$$FeS_2 + 2O_2 = FeSO_4 + S^\circ \tag{4}$$

ii.   in sulfuric acid solutions

$$2FeS_2 + 2H_2SO_4 + 3O_2 = Fe_2(SO_4)_3 + 3S^\circ + 2H_2O \tag{5}$$

iii. in nitric acid solutions

$$3FeS_2 + 18HNO_3 = Fe_2(SO_4)_3 + Fe(NO_3)_3 + 3H_2SO_4 + 15NO + 6H_2O \tag{6}$$

$$2FeS_2 + 10HNO_3 = Fe_2(SO_4)_3 + H_2SO_4 + 10NO + 4H_2O \tag{7}$$

iv.   in hydrogen peroxide solutions

$$FeS_2 + 7.5H_2O_2 = Fe^{3+} + 2SO_4^{2-} + H^+ + 7H_2O \tag{8}$$

v.   in highly acidic solutions

$$FeS_2 + 7.5H_2O_2 + H^+ = Fe^{3+} + 2HSO_4^- + 7H_2O \qquad (9)$$

(vi) in presence of ozone
Direct oxidation

$$FeS_2 + O_3 + H_2O + 2 O_2 = FeSO_4 + H_2SO_4 \qquad (10)$$

$$FeS2 + 7/3O_3 + H_2O = FeSO_4 + H_2SO_4 \qquad (11)$$

Indirect oxidation

$$2/3O_3(g) = O_2(ac) \qquad (12)$$

$$FeS_2 + 7/2 O_2 + H_2O = FeSO4 + H_2SO_4 \qquad (13)$$

The oxidation of sulfide ores by ozone can occur by dissolution of sulfide species and the formation sulfate ion, as suggested by Elorza et al. Then, the global reaction of the pyrite oxidation in the presence of ozone can be described as follows:

$$FeS_2 + 2/3O_3 + 5/2O_2 + H_2O = Fe^{2+} + 2SO_4^{2-} + 2H^+ \qquad (14)$$

In the case of chalcopyrite, the acid leaching in presence of Fe occur according to the following reaction:

$$Cu\ FeS_2 + 4Fe^{+3} = Cu^{+1} + 5Fe^{+2} + 2S° \qquad (15)$$

According to Havlik et al. (7, 9), the global reaction of chalcopyrite under the action of O_3 can be represented by:

$$3CuFeS_2 + 8O_3 = 3CuSO_4 + 3FeSO_4 \qquad (16)$$

2.3 Kinetics

In hydrometallurgy, most of leaching process follows the kinetic models for heterogeneous solid/liquid reactions, known as shrinking core models (SCM), as showed in Figure 3: the SCM controlled by chemical reaction and the SCM controlled by diffusion trough the solid product layer (Habashi, 1999; Levenspiel, 1999; Sohn and Wadsworth, 1986). A third model, the stochastic model for control by chemical reactions on the non-reacted particle surface (Ciminelli and Osseo-Assare, 1995) is considered.

In the mentioned models, the fraction of iron reacted at any time t, can be predicted from the following Equation.

1. Shrinking core model controlled by the chemical reaction

$$kt = 1 - (1 - x)^{1/3} \qquad (17)$$

Where, x is the fraction of iron reacted and can be calculated from the following relation:

$$x = \frac{C}{C_0} \qquad (18)$$

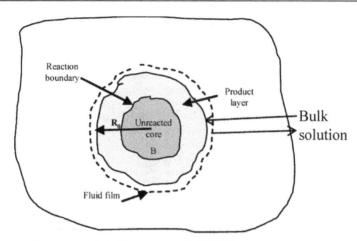

Fig. 3. Squematic diagram of shrinking core model (SCM).

And κ, is the apparent rate constant, and can be calculated from the following relation:

$$k = \frac{k_s C_A}{R_0 \rho} \qquad (19)$$

Where, k_s is the rate constant of the reaction, ρ is the density of the FeS_2 ore, R_0, is the radius of the un-reacted particle, and C_A, is the reactive concentration in the solution. The above equations are applied to mono-sized particles, thus the average size of a narrow fraction of particles can be used in the kinetic model.

2. Shrinking core model controlled by the diffusion of the reagents or dissolved species through the layer of solid reaction products, the fraction of iron reacted at any time t can be predicted from the following equation.

$$kt = 1 - \frac{2}{3}x - (1 - x)^{2/3} \qquad (20)$$

Where κ can be calculated from the following relation:

$$k = \frac{2DC_A}{R_0^2 \rho} \qquad (21)$$

Where, D, is the diffusion coefficient of the iron species.

3. Stochastic model. It takes into account the heterogeneity of solid minerals by introduction a stochastic distribution for the rate constant. Then, the rate constant, k_s from the shrinking core model is transformed into a variable that changes with time or conversion, according to following relation:

$$k_s(X) = 2\, k_0(1\text{-}X) \qquad (22)$$

where, $k_s = k_{max}/2$

According to Ciminelli and Osseo-Assare (1995), the resulting equation has the following expression:

$$kt = (1-x)^{-2/3} - 1 \qquad (23)$$

$$k = \frac{4k_sC_A}{\rho R_0} \qquad (24)$$

2.4 Costs

Ozone can be produced by many ways. There are more than 700 patented ways of ozone production. But commercially three most popular methods are being used: a) The UV method of ozone production, b) The plate types Corona ozone production and c) The tube types Corona ozone production (Baratharaj, 2011).

For UV method ozone production, it is therefore necessary to utilize a short wavelength ~185nm. In theory, the yield of O_3 from 185nm UV light is 130g/kWh of light. As lamp efficiencies are so low, ~1%, the production per kWh from the power source is greatly reduced. In practice, with the present state of development, UV lamps can only produce about 20g O_3/kWh of ozone when using oxygen as the feed gas (Smith, 2011).

Ozone production by electrical discharges has been, and remains, the most commercially viable method. Essentially a corona is characterised by a low current electrical discharge across a gas-filled gap at a relatively high voltage gradient. The amount of ozone produced in a given corona ozonator design is relative to the concentration of oxygen in the gas feeding the corona. Basically, the more oxygen in, the more ozone out. In general, ozone concentrations of 1-3% using air, and 3-10% using oxygen can be obtained. The amount of energy applied to the gas gap between the electrodes is critical to the concentration of ozone produced. It is a combination of the voltage and frequency that results in a given energy input. Typically, voltages of between 7 to 30 kV are used with frequencies ranging from mains supply of 50 or 60 Hz, medium up to 1000 Hz, and high up to 4000 Hz. Then, the net effect is that less power is consumed to generate a given quantity of ozone as the oxygen concentration increases: Oxygen ~ 5 to 8kW/kg; Air ~ 15 to 18kW/kg (Smith, 2011).

Although there are no reports on the cost of using ozone in industrial mining applications, Botz et al (2000) reported a pilot scale study for oxidation of cyanide in mining effluents. In this case, the cost of using ozone operation was U.S. $ 0.97 per kg cyanide removed, compared with SO_2/air and chlorination methods, $ 1.35 and $ 1.67, respectively.

Therefore, economical feasibility of the use of ozone in oxidation of sulphide minerals is possible, taking into account the amount of ozone used and the energy cost of production.

3. Cases of study

3.1 Oxidation of sulphide ores-containing gold and silver

Cyanidation is the most aqueous leaching process used to extract gold and silver. However, it has some disadvantages when precious metals are encapsulated in matrixes of iron sulphide minerals, such as arsenopyrite and pyrite (Shoemaker, 1990). In this case, the minerals receive an oxidation pretreatment (as oxidation roasting, chemical oxidation under pressure or biological oxidation) to facilitate gold and silver extraction by cyanide solution (Weir and Berezowsky, 1986; Chen and Reddy, 1990; Burbank et al., 1990). An alternative to these methods is the use of ozone, which increases the oxidation potential and the oxygen content of solution during cyanidation (Haque, 1992; Roca et al., 2000; Salinas et al., 2004; Elorza et al., 2006; Carrillo et al., 2007). Ozone can create favourable oxidation conditions for sulfide minerals in aqueous mediums. According to the Eh – pH diagram shown in Figure 1 and Eq. 10 to 13, the sulfide species, such as pyrite or pyrrhotite, can be oxidized to sulfate in oxidant conditions and within a pH range from 2 to 14, making it possible to obtain solids or solutions. In both cases, the product formed during the oxidizing reaction of pyrite with ozone permits favorable conditions to the contact of cyanide and oxygen with precious metals containing in the ore, thus increasing the efficiency of the cyanidation process and, on the other hand, the sulfur oxidized to sulfate will no longer react with cyanide to form the thiocyanate ion SCN-, one of the causes of the increased consumption of cyanide during cyanidation.

Although in the mining industry, especially in the case of extraction of precious metals, the use of ozone has not been much discussed, laboratory experiments indicate that ozone may be a valid alternative for resolving or surmounting the disadvantages of the already mentioned cyanidation process. In the case of refractory minerals, there have been reports of increases in the recovery of gold and silver which vary from 25% to more than 100% for both cases, as well as a significant reduction in the time of cyanidation (Salinas et al. 2004; Elorza et al., 2006). With non-refractory pyritic minerals, the results obtained have shown that pre-treatment with ozone not only permits a greater extraction of gold during cyanidation, but also causes less cyanide consumption.

Table 1 shows gold and silver composition of samples of pyrite containing gold and silver, with a size distribution of 75% -75 μm. The detailed process for experimental tests was previously reported (Carrillo et al., 2007, 2011). Experiment included a pretreatment (before cyanidation) with ozone directly in mineral slurry at pH of 6. Subsequently, solid sample was treated for 48 h under conventional cyanidation conditions. Table displays the amount of metal recovered from the cyanidation process. It is evident that ozone pretreatment increased dissolution of gold in cyanidation, particularly for the sample with the highest gold composition. For samples A and B, dissolution value increased significantly with ozone, and in sample C is interesting to note that it is still possible to recover gold by cyanidation method. Table 1 shows silver dissolved percentage during cyanidation process. When ozone pretreatment is carried out, the amount of silver dissolved increased to 82, 83 and 75 %, respectively. The increase in samples A and B was about 15%, but sample 3 showed a very significant increment in extracted silver. The mineralogy of this metal could explain the difference: the sample was probably in form of argentite, a silver sulfur that is not extracted during cyanidation process. Previous results suggested that ozone introduced in the slurry, chemically reacts with pyrite´s sulfur increasing the oxidation potential of the slurry. In a previous work, we have shown that ozone treatment leads to partial oxidation of sulphide minerals and to sulfate ion formation, specifically in oxidation of sulfur (Carrillo et al., 2007). Improvement in gold and silver recovery from ore with ozone pretreatment indicated that reaction intermediate products promote the conditions for cyanide diffusion to the precious metals in the subsequent cyanidation process.

	Sample A		Sample B		Sample C	
	Au	Ag	Au	Ag	Au	Ag
Initial concentration, g/ton	4.5	435	6.7	326	0.3	34
Extraction, %, Cyanidation without O_3 pre-treatment	84.69	71.51	76.25	71.24	70.20	16.67
Cyanide consumption, kg/ton, without O_3 pre-treatment	5.8		3.12		3.6	
Extraction, %, Cyanidation with O_3 pre-treatment	91.86	81.99	97.39	83.33	61.70	75
Cyanide consumption, kg/ton, with O_3 pre-treatment	1.36		1.2		1.2	

Table 1. Chemical Assay of pyrite samples containing gold and silver, used in dissolution of precious metals by cyanidation (48 h) with and without ozone pre-treatment.

Table 1 also shows the consumption of cyanide during the cyanidation of the samples, with and without pre-treatment. It can be seen that, after pre-oxidation, the consumption of cyanide was high. 5.8, 3.12 and 3.6 kg/ton. For the same samples, the consumption of cyanide decreased considerably with the pre-oxidation, achieving a significant saving compared to untreated samples. According to the results obtained, ozone treatment before cyanidation permits partial oxidation of sulphide minerals, specifically the oxidation of sulphur. The conditions necessary for these reactions would be acidic pH less than 6, and oxidants conditions according to Figure 1. These conditions, pH of 6 in the slurry, can be maintained during the ozone pre-treatment test. The ozone introduced in slurry reacts by chemical reaction with sulfur of pyrite or by ozone decomposition at oxygen, increasing the oxidation potential of the slurry. The fact that recovery of gold and silver improved with oxidation pretreatment of the ore indicates that the obtained product of reaction contributes to creating better conditions for the diffusion of cyanide to the precious metals. However, the low consumption of ozone in the tests permits one to suppose that the principal function of ozone is increase the oxidation potential of slurry, and obtain better conditions for partial pyrite oxidation. In addition, treatment with ozone permits increasing the content of soluble oxygen in the ore slurry, the oxygen that is the product of the decomposition of the ozone during its reaction with the ore. The presence of more oxygen in the slurry permits better conditions for the complexes formation during the cyanidation reaction, since it has been reported already that cyanidation is an electrochemical reaction produced by the cathodic reaction of the oxygen in the surface of the metal, which permits the anodic dissolution of the precious metals in order to achieve the complete cyanidation (Habashi, 1970; Parga et al., 2003). On the other hand, the oxidation of sulfur to sulphate or sulfate ion prevents sulfur species from reacting with cyanide to form thiocyanate, reaction which, as was mentioned above, is the main cause for the consumption of cyanide with this type of sulfide ores. Therefore, the consumption of cyanide decreases, thus increasing the efficiency of the process and decreasing its cost.

3.2 Oxidation of sulphide copper minerals

Sulphide copper minerals, such as chalcopyrite (CuFeS2), are the most abundant copper-bearing minerals, and represent approximately 70 % of the world's known copper reserves

(Davenport et al., 2002). Chalcopyrite is also the most stable of copper minerals due to its structural configuration (face-centered tetragonal lattice) and, consequently, the most refractory for aqueous extraction processing.

Industrially, copper ore leaching is almost always accomplished by diluted sulfuric acid medium and ferric sulphate, which are low-cost reagents and could be regenerated when ores are lixiviated. Several studies have been conducted to optimize the process conditions and to explain the basics of chalcopyrite leaching process. Thus, it has been suggested that a layer of elemental sulfur is formed on the external surface. The type of sulfur layer formed on the surface, according to Eq. 15, depends upon the reagents used, as well as on the process conditions (i.e., temperature and agitation); importantly, this layer inhibits the dissolution of the chalcopyrite, thus reducing the overall leaching rate and the process efficiency. Several approaches have been recommended to accelerate the chalcopyrite dissolution. However, there is an increasing interest in optimizing the aqueous extraction process for copper production due to the negative environmental impact caused by chemical reagents used (Shijie, 2005; Peacey et al, 2003). Although leaching of copper ores is carried out in diluted sulfuric acid medium and ferric sulfate as oxidant (Ukasik and Havlik, 2005; Antonijevic and Bogdanovic, 2004), low-cost reagents, different approaches have been suggested to increase chalcopyrite rate dissolution. The most common is to increase process temperature, but this implies higher energy requirements. Another suggested alternative is the use of strong oxidants such as ozone (Havlik et al., 1999), hydrogen peroxide (Antonijevic et al., 2004) and manganese nodules (Havlik et. al., 2005).

On the other hand, the use of ferric ion (Fe^{+3}) to dissolve copper has an economic constraint and, therefore, Fe^{+3} has to be regenerated. This could be accomplished by oxidizing ferrous ions (Fe^{+2}) with air and oxygen, although this step is usually very slow in acid medium. Pressure oxidation is an alternative for this oxidation process, but it is only applied on concentrated ores . Nevertheless, a thermodynamic point of view, the most effective way to improve the process efficiency is to eliminate the formation of any sulfur layer on the chalcopyrite surface and, at the time, regenerate Fe^{+3}, and this is possible under strong oxidation conditions.

In the leaching of a mixed ore, the high redox potential required in an acidic medium to avoid the sulfur layer formation can be met by using ozone (O_3) as oxidizing agent.

Havlik et al. (9) studied the leaching kinetics of a chalcopyrite concentrate in 0.5 M sulfuric acid ($H2SO4$) solutions, using O_3 as oxidizing agent, in the range of 4 to 75°C. Under the studied conditions, the reaction showed parabolic kinetics. No evidence of the formation of an elemental sulfur layer, or any other product layer, was found. The authors also indicated that the overall reaction rate was controlled by diffusion of O_3 in the interface solid-liquid; in addition, they reported that solubility of ozone decreases as the temperatures rises above 40 ° C, thus limiting the beneficial effect of a temperature increase.

Carrillo et al. (2010) mentioned that, according to equation 1, as the reaction takes place, a significant amount of ferrous ion (Fe^{+2}) is formed. The continuous addition of O_3 into the solution favors the oxidation of Fe^{+2} to Fe^{+3}, according to following equation.

$$6Fe^{+2} + O_3 + 6H^+ = 6Fe^{+3} + 3H_2O \tag{25}$$

The occurrence of reaction 24 in preferential conditions such as the reaction rate is enhanced (i.e., large O_3 concentration, low pH) might cause an increase in (Fe^{+3}), which in turn should favor reaction 1 and, in consequence, copper dissolution.

Therefore, a possible mechanism for chalcopyrite dissolution in the presence of Fe^{+3} and O_3 is that Fe^{+3} quickly react with the mineral surface to produce copper ions, Fe^{+2} ions and also an sulfur compounds (sulphate) layer on the surface. Then, Fe^{+3} and Fe^{+2} ions must diffuse through this layer to continue with the dissolution process. In addition, O_3 must diffuse from the gas bulk to the solution and to the interface of chalcopyrite particles and react with chalcopyrite (equation 2) and Fe^{+2} ions (equation 6). This last step takes place in the solution and might be faster, leading to the formation of more Fe^{+3} in the solution.

Figure 4 show copper profiles based on a Taguchi L9 experimental design. The figures show the main effects, as determined with a S/N (Signal/Noise) ratio, which is based on the concept of the "greater-the-better", was used to characterize the response (amount of copper extracted). The S/N ratio was defined as:

$$S/N \text{ ratio} = -10\log(MSD) \qquad (26)$$

where, the MSD (mean-square deviation) was calculated by:

$$MSD = 1/n \sum yi^2 \qquad (27)$$

Where, n was number of tests, and yi was the value of Cu extracted (%) obtained from the ith test., as a function of the amount of copper extracted. Accordingly, figure shows that

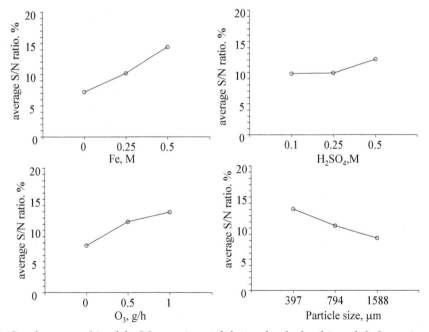

Fig. 4. One factor graphic of the L9 experimental design for the leaching of chalcopyrite.

under the studied conditions, (Fe^{3+}) is the most important factor during the recovery of copper by chemical dissolution of chalcopyrite. Results also indicated that, within the analyzed range, (H_2SO_4) had no effect on the amount of copper extracted. An increase in the levels of (Fe^{3+}) and O_3 concentration from first level to second level, and from second level to third level, resulted in a increase in the amount of copper extracted. Similar results were found for a decrease in the levels of particle size. However, when the levels of (H_2SO_4) were increased, no significant effect was observed in the response.

The results showed in figure suggest that Fe^{+3} react at the interface increasing copper the dissolution. Obviously, reduction of the particle size increases the liberation grade of chalcopyrite and the reaction surface area, thus exposing a larger fraction of the copper mineral, and promoting a better contact between the metal and the chemical agents (Fe^{3+}, O_3 and H_2SO_4) for faster dissolution. In addition, the strong oxidant ion condition is enhanced when O_3 it used. This beneficial effect is found for all (Fe^{+3}) used. Therefore, a possible mechanism for chalcopyrite dissolution in the presence of Fe^{+3} and O_3 is that Fe^{+3} quickly react with the mineral surface to produce copper ions, Fe^{+3} ions and also an sulfur compounds (sulphate) layer on the surface. Then, Fe^{+3} and Fe^{+2} ions must diffuse through this layer to continue with the dissolution process. In addition, O_3 must diffuse from the gas bulk to the solution and to the interface of chalcopyrite particles and react with chalcopyrite and Fe^{+2} ions. This last step takes place in the solution and might be faster, leading to the formation of more Fe^{+3} in the solution. Then, the increased concentration of Fe^{+3} might promote the copper dissolution process, but diffusion of $Fe+3$ ion in sulfur compounds layer could be slower, and thus gradual stop of the overall rate of copper extraction.

Based on results, the effect of adding Fe^{+3} and O_3 is favourable for small particle sizes. For larger particles, only (Fe^{+3}) seemed to affect copper dissolution since an increase in O_3 had no beneficial effect. It has to be considered that for larger particle sizes, there is less chalcopyrite exposed to the reagents and, therefore, the presence of a single oxidant is sufficient to promote copper dissolution. Due to the relatively small liberation of chalcopyrite, the oxidation of Fe^{+2} with O_3 is not a limiting step to promote copper dissolution. Finally, the results obtained here suggest that only minimal quantities of acid in the solution are required for the dissolution of copper, just enough to prevent hydrolysis and precipitation of Fe^{+3} by OH^-.

3.3 Iron sulphide oxidation in coal

Another process is ozone application for iron sulphide oxidation in coal, one of the most important fossil fuels used for energy production. However, due to its nature, coal requires a cleaning stage based on physical methods before its use to meet air pollution regulations (Apenzaller, 2006), but organic sulfur and syngenetic pyrite is removed with low efficiency to the required level (Ozbayoglu, 1998). Previous to the combustion, coal cleaning techniques based in physical methods are extensively used, but are less efficient to remove organic sulfur and syngenetic pyrite ($FeS2$). Syngenetic pyrite is one of the two forms of pyritic sulphur, found as a very fine and highly disseminated mineral in coal, which makes it difficult to separate by conventional cleaning process (Baruah and Khare, 2007; Pysh`yevl et al., 2007; Li and Cho, 2005; Ayha et al., 2005; Baruah et al., 2006).

Many studies have been realized to explore the pyrite dissolution by oxidants aqueous media. For this purpose, various oxidizing agents such as oxygen, hydrogen peroxide, ferric sulfate, ferric chloride, potassium permanganate, perchloric and nitric acids have been used to oxidize pyrite (Elliot, 1978; Bonn and Heijnen, 2001; Borah, 2006; Kawatra and Eisele, 2001; Antonijevic et al., 2003; Karaca et al., 2003; Mukherjee and Srisvastava, 2004). Previous work reported author indicate that hydrogen peroxide, ozone, and combined ozone-hydrogen peroxide in acid medium can to help to the pyrite removal (Davalos et al., 2009; Carrillo et al., 2009).

The oxidation of pyrite in an acid medium has been extensively studied and documented due to its importance in sulfur processing. This anodic process is recognized as a complex method that involves chemical and electrochemical equilibrium. Exactly as Chander et al. have summarized, pyrite oxidation processes have been classified into two mechanisms: (a) the preferential release of iron ions from pyrite and (b) the preferential release of oxysulfuric species. In the first process, the outer reacted layer of pyrite has been identified as elemental sulfur (S°), "polysulfide," or "metal deficient," which corresponds to the theory proposed by Buckley et al. In the second process, the sulfur in pyrite oxidizes to sulfates or thiosulfates, leaving a reacted layer composed of iron hydroxides.

The use of ozone for pyrite removal in coal and its effect in different conditions were investigated by Dávalos et al. (2009) and Carrillo et al.(2010). The use of ozone and its effect in different conditions were investigated. These works were based on an experimental design with the following parameters and levels: type and concentration of reagents (NaOH, HCl, HNO_3 and H_2SO_4; 0.3, 0.8 and 1.3 M; and distilled water) and presence and concentration of O_3 (0, 0.16 and 0.33 L·g/hr). The main factors affecting the FeS_2 dissolution were determined by analysis of variance (ANOVA). Table 2 shows the main effect ANOVA of the results as a function of the amount of Fe extracted at 90 minutes of treatment. According to the table, ANOVA shows that, under the studied conditions, the type of acid and concentration of O3 employed are the most important factor of the chemical dissolution of pyrite, followed of the reagents concentrations. Test results confirms that the maximum pyrite dissolution is reached when sulphuric acid is used. Increasing the acid concentration (average of the different acids), does not clearly affect the response of these curves. An increase in the levels of factors O_3, results in an increase in the mean value. Based on this, the results indicate a qualitative way to know the behaviour of pyrite dissolution at different combinations of chemicals (aqueous medium, concentration, oxidants).

Source term	DF	Sum of squares	Mean square	F-rato	Prob level	Power ($\alpha = 0.05$)
Reagent type	4	112.83	28.21	0.88	0.5642	0.1122
Reagent concentration	5	235.99	47.20	1.48	0.3979	0.1580
Ozone	2	123.61	61.80	1.93	0.2890	0.1804
S	3	95.99	31.99			
Total (adjusted)	14	1664.24				
Total	15					

Table 2. ANOVA for experimental design used in pyritic removal of coal samples.

In FeS$_2$ dissolution, the previously assumption that the extent of the chemical reaction rate at the interface is similar than that of the diffusion, can be precise. The presence of a strong oxidant conditions prevents the formation of the aforementioned product layer of elemental sulphur due to the high anodic potential achieved, or, at least, the electrochemical conditions influence the layer texture, favouring the formation of a porous cover of product.

The dissolution of pyrite proceeds into a coal matrix, thus the diffusion through the porous in the coal could limit the flux of the reagents and products to or from the reactive layer of pyrite. In addition, to the diffusion, the reagents (alkali and acid) could modify the chemical properties of coal, therefore a consumption of oxidative reagents would occur, thus leading the process to the chemical control. The diffusion and the reagents consumption, could explain the fit of the results to the diffusion and chemical controlling stage. Table 3 shows the kinetics data of pyrite dissolution form coal with the different reagents. This table shows that the SCM for product layer diffusion control, describes well the experimental data with respect to the others. Then, the evidence of kinetics behavior of the experimental data indicates that the reaction can be controlled by diffusion through a layer or film conformed by the surrounding surface coal.

Model	HNO$_3$	HCl	NaOH	H$_2$SO4	H$_2$O
$kt = 1 - (1-x)^{\frac{1}{3}}$	0.91	0.68	0.83	0.78	0.85
$kt = 1 - \frac{2}{3}x - (1-x)^{\frac{2}{3}}$	0.78	0.53	0.52	0.57	0.72
$kt = (1-x)^{-\frac{2}{3}} - 1$	0.81	0.56	0.53	0.63	0.74

Table 3. Correlation coefficient (r2) obtained from the kinetics models for FeS$_2$ dissolution.

In order to compare the results of O$_3$ treatment (leaching) with washability and flotation tests, different coal samples were treated, as shown in Table 4, showing the sulfur content of each product obtained in each test. The O$_3$ treatment was carried out using H$_2$SO$_4$ as aqueous medium solution. The results indicate that sulfur is largely free, so from the first separation in dense medium (washability test), with density 1.3, can get a 27% removal of sulfur. This indicates a ratio of sulfur released that has a higher density than coal. Significantly, the decrease depends on the initial sulfur content in the sample, but the trend continues in the different test.

In subsequent steps of washability separation, sulfur removal rate of 27% is maintained, except the last, which is 18%. Clearly, the washability test is cumulative, so that the sulfur content is relative to the sample proportion remaining in each stage. In the flotation to recover fine coal used, it is possible to obtain low sulfur content, although the removal with respect to the original sample is on average 11%. Although a smaller particle size (100% - 600 mesh Tyler), and therefore possibly a greater release of sulfur, it is also possible that a proportion of pyrite released to interact with the collector or pyrite fines being swept by the foam. On the other hand, it is interesting to note that the sulfur content obtained by the O$_3$ treatment of sample M1-CAO (sample obtained of coal mixed samples, named coal all-in-one) is similar to the first separate density 1.3 (25% removal) and a sample washed coal and may further reduce the sulfur content of such leaching, reaching up to 8% sulfur removal.

Sample	Initial Sulfur content, %	Washability tests at different density medium					Flotation test
		d= 1,3	d=1,4	d=1,5	d=1,6	bottom 1,6	
Block VII	1.35	0.86	0.76	0.88	1.16	1.15	1.11
Block VI	1.17	0.96	0.98	1.07	1.11	1.72	1.1
Mine 6	1.22	0.79	0.75	0.75	0.95	0.9	1.05
Mine 5	1.09	0.79	0.79	0.79	0.75	0.97	1
Mine 3	1.02	0.87	0.79	0.77	0.79	1.24	0.95
Muestras	M1	After O_3 treatment					
M1-CAO	1.17	0.88					
M1-washed	0.87	0.78					

Table 4. Total sulfur, %, for the diferent removal sulfur test.

To confirm the previous results in industrial sample, CAO (labeled M2) samples were obtained from different steps (P) of industrial washing plant. These samples were treated with oxidant leaching tests, which were repeated 4 times. The objective of this procedure was to determine the reduction of sulfur, also determine the degree of repeatability and reproducibility (R) of the tests.

Figure 5 shows the initial sulfur content in each sample (hollows circles), as well as the average, minimum and maximum sulfur analysis obtained for each sample treatment. The error bar indicates the range of results from 4 trials (repetitions) performed on each sample.

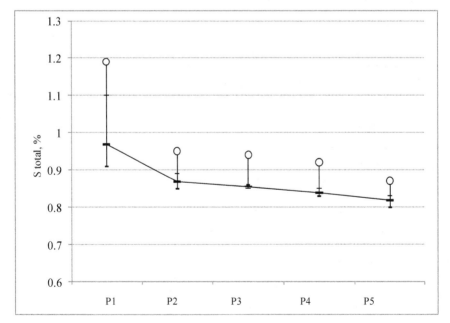

Fig. 5. Total sulfur content (%) in samples obtained from different washing steps (P1 – P5), treated by O_3 leaching. Initial sulfur content in original sample = 1.19 %.

Results indicate that removal of sulfur is greater if the initial sulfur content is higher, although the variability obtained is also high. For example, with P1 step coal, an average 18 % sulfur removal was obtained. When the initial sulfur is lower, it also reduces the variability in treatment result, but the removal percentage is lower, between 8 and 9%. This may be due to the particle size distribution that exists in the wash steps, which consists of steps processes using cell dense media (box Daniels), hydro-cyclones and spirals. The particle size is related to the degree of liberation of sulfur (pyrite). Then, a narrower size distribution, the variability is lower.

The results indicate that ozone is the stronger oxidant to remove the pyrite and its dissolution is enhanced by sulphuric acid medium. The analysis of the results with O_3 as oxidising agent using ANOVA and multiple lab-test, was applied in order to support that diffusion of oxidant agents is the rate controlling step of the overall process: O_3 diffusion from the gas bulk to the dissolution to react with pyrite and diffusion of Fe^{3+} through the layer formed in the boundary surface pyrite and coal particles. The results clearly showed that ozone contributed to the use of lower concentrations of H_2SO_4, with the consequently economical savings.

4. Conclusion

Sulfur is the element that, in copper, gold and silver cases, inhibits metal dissolution present in minerals. On the other hand, the removal of sulfur is greatly needed in case of coal, due to emissions of SO_2 gas during coal fire-combustion. In this Chapter, oxidative hydrometallurgy of different sulphide minerals using ozone has been analyzed. Results show that the treatment with ozone increases sulfur oxidation in the different cases.

Pyrite containing gold and silver results suggests that the treatment with ozone before cyanidation process improves the recovery of gold and silver and reduces the consumption of cyanide. The results obtained from low metal value content sample indicate that is still possible to recover more than half of the gold and silver content by means of cyanidation. The results indicate a significant reduction in the consumption of cyanide which permits the recovery of these metals with less cost. But, the determination of the cost of ozone consumption in order to evaluate its use is very important.

In the case of chalcopyrite, the approach suggested in this work using experimental design with Taguchi L9 design matrix and statistical analysis to study the acid leaching of chalcopyrite with ozone and ferric ions was helpful to optimize the experimental conditions and to rationalize the results. This approach may be used to decrease development costs and ultimately operational costs of copper extraction from low-grade chalcopyrite.

For sub-bituminous coal, the dissolution of pyrite from in different medium using ozone was studied. The results, using ANOVA, indicate that ozone is the stronger oxidant to remove the pyrite and its dissolution is enhanced by sulphuric acid medium, the results support that diffusion of oxidant agents is the rate-controlling step of the overall process: O_3 diffusion from the gas bulk to dissolution to react with pyrite and diffusion of Fe^{3+} through the layer formed in the boundary surface pyrite and coal particles.

Thus, the use of ozone can be a promising auxiliary agent in the actual process of obtaining metals and coal with the following advantages: 1) decreasing operational costs of low-grade chalcopyrite leaching, 2) increasing gold and silver recovery in cyanidation and decrease cyanide consumption, 3) decreasing the sulfur-containing coal since coal cleaning plants, to be used as clean in energy generation and iron and steelmaking.

5. Acknowledgment

Authors thanks to CGEPI-UAdeC, PROMEP-SEP and CONACYT for financial support (Projects 67039).

6. References

Antonijevic, M., Bogdanovic, M., (2004), Investigation of the leaching of chalcopyritic ore in acidic solutions, *Hydrometallurgy* 73, 3-4, 245-256.

Antonijevic, M. M., Jankovic, Z. D., Dimitrijevic, M. D., (2004), Kinetics of chalcopyrite dissolution by hydrogen peroxide in sulphuric acid, *Hydrometallurgy* 71, 329-334.

Antwerp, W. P. V. and Lincoln, P. A., (1987), Precious Metal Recovery using UV ozone, US Patent 4642134:8 (1987).

Apenzaller, T., (2006), The coal paradox, *National Geographic*, March, 99-103.

Baratharaj, V., (2011), How to evaluate and select an ozone generator, in *http://www.otsil.net/articles*.

Baruah, B., Saika, B., Kotoky, P., Rao, P., (2006), Aqueous leaching on high sulfur sub-bituminous coals, in Assam, India, *Energy & Fuels*, 20, 4, 1550-1555.

Borah, D., (2006), Desulfurization of organic sulfur from a subbituminous coal by electron transfer process with K4(Fe(CN)6), *Energy & Fuels*, 20, 1, 287-294.

Burbank, A. Choi, N. and Pribrey, K. , (1990), Biooxidation of Refractory Gold Ores in Heaps, *Proc. Advances in Gold and Silver Processing*, Reno, NV, USA, 151-159.

Carrillo, F., Soria, M., Martínez, A., Gonzalez, A., (2007), Ozonation pretreament of gold-silver pyritic minerals, *Ozone Sci. & Eng.*, 29, 307-313.

Carrillo, F. Davalos, A.., Soria, M., Pecina, T., (2009), Coal desulfurization in oxidative Acid media using hydrogen peroxide and ozone: A kinetic and statistical approach, *Energy & Fuel*, 23, 3703–3710.

Carrillo-Pedroza, F. R., Sánchez-Castillo, M., Soria-Aguilar, M., Martínez-Luévanos, A. and Gutiérrez, E., (2010), Evaluation of Acid Leaching of Low Grade Chalcopyrite Using Ozone by Statistical Analysis, *Canadian Metallurgical Quarterly*, 49, (3), 219-226

Davalos, A., Pecina, T., Soria, M. and Carrillo, F., (2009), Kinetics of Coal Desulfurization in An Oxidative Acid Media, *Int. J. of Coal Prep. and Utilization*, 29 (3), 152 – 172.

Chen, B. and Reddy, R. G., (1990), Roasting Characteristic of Refractory Gold Ores, Proc. Advances in Gold and Silver, Reno, NV, USA, 201-214.

Deng, T., (1992), Chemical Oxidation of Iron Sulphide Minerals for Metals Recovery", Min. Process. and Extract. Metall. Rev., 10, 325-345.

Elliot, R. C.,(1978), *Coal Desulfurization Prior to Combustion*, Ed. Noyes Data Corporation.

Elorza-Rodríguez, E., Nava-Alonso, F. Jara, J. and Lara-Valenzuela, C., (2006), Treatment of Pyritic Matrix Gold–Silver Refractory Ores by Ozonization–Cyanidation, *Minerals Engineering*, 19,1 56-61.

Habashi, F. (1999), *Kinetics of Metallurgical Process*, Métallurgie Extractive Québec, Québec City Canada.

Haque, K. E. (1992), The Role of Oxygen in Cyanide Leaching of Gold Ore, *CIM Bulletin*, 85, 963, 31-38.

Havlik, T., Dvorscikova, J., Ivanova, Z., Kammel, R., (1999), Sulphuric acid chalcopyrite leaching using ozone as oxidant, *Metall*. 53 (1-2), 57-60.

Havlik, T., Laubertova, M., Miskufova, A., Kondas, J., Vranka, F., Extraction of copper, zinc, nickel and cobalt in acid oxidative leaching of chalcopyrite at the presence of deep-sea manganese nodules as oxidant, *Hydrometallurgy*, 77, 1-2, 51-59 (2005).

Hill, A. G. y Rice, R. G., (1982), Historical Background, Properties and Applications, *Handbook of Ozone Technology and Applications, Vol. 1*, (Edited by Rip G. Rice and Aharon Netzer, Ann Arbor Science Publishers, 1982), 105-142.

Kawatra, S. K. y Eisele, T. C., (2001), *Coal Desulfurization*, Taylor & Francis.

Ozbayoglu, G., (1998), *Mineral Processing and the Environment*, Ed. Gallios, G. P. and Matis, K. A., Kluwer Academic Publishers, pp. 199 –221

Peacey, J., Guo, X. J., Robles, E., (2003), Copper hydrometallurgy – curren status, preliminary economics, future directions on positioning versus smelting, in *Copper 2003* Vol. VI, , CIM-MetSoc, 205-222

Rice, R. G., (1997), Applications of Ozone for Industrial Wastewater Treatment – A Review, Ozone; Science and Engineering, 18, (6): 477-516.

Roca, A. Cruells, M. and Viñals, J., (2000), Aplicaciones del Ozono en los Sistemas Hidrometalúrgicos, *Proc. X International Congress in Extractive Metallurgy*, México, 22 – 32.

Salinas, E. Rivera, I. Carrillo, R. Patiño, F. Hernández, J. Hernández, L. E., (2004), Mejora del Proceso de Cianuración de Oro y Plata, mediante la Preoxidación de Minerales Sulfurosos con Ozono, *Rev. Soc. Quím. Méx.*, 48, 225-356 (2004).

Shijie, W., (2005), Copper leaching from chalcopyrite concentrates, JOM, 57, 7, 48-52.

Smith, Wayne, (2011) Principles of Ozone Generation, in *http://www.watertecengineering.com*.

Shoemaker, R. S., (1990), Refractory Gold Processing, *Proc. Advances in Gold and Silver*, Reno, NV, USA, 113-118, 1990.

Ukasik, M., Havlik, T., (2005), Effect of selected parameters on tetrahedrite leaching by ozone", Hydrometallurgy 77, 1-2, 139-145.

Weir, D. and Berezowsky, R., (1986), Refractory Gold: The Role of Pressure Oxidation", *Proc. International Conference on Gold*, Johannesburg, SA, 275-285.

Part 3

Modelling of the Refining Processes in the Production of Steel and Ferroalloys

4

Modelling of the Refining Processes in the Production of Ferrochrome and Stainless Steel

Eetu-Pekka Heikkinen and Timo Fabritius
University of Oulu,
Finland

1. Introduction

In stainless steel production - as in almost any kind of industrial activity - it is important to know what kind of influence different factors such as process variables and conditions have on the process outcome. In order to productionally, economically and ecologically optimize the refining processes used in the production of stainless steels, one has to know these connections between the process outcomes and the process variables. In an effort to obtain this knowledge, process modelling and simulation - as well as experimental procedures and analyses - can be used as valuable tools (Heikkinen et al., 2010a).

Process modelling and optimization requires information concerning the physical and chemical phenomena inside the process. However, even a deep understanding of these phenomena alone is not sufficient without the knowledge concerning the connections between the phenomena and the applications because of which the process modelling is carried out in the first place. The process engineer needs to seek the answers for questions such as: What are the applications and process outcomes (processes, product quantities, product qualities and properties, raw materials, emissions, residues and other environmental effects, refractory materials, etc.) that need to be modelled? What are the essential phenomena (chemical, thermal, mechanical, physical) influencing these applications? What variables need to be considered? What are the relations between these variables and process outcomes? How these relations should be modelled? (Heikkinen et al., 2010a)

The purpose of this chapter is to seek answers to these questions in the context of ferrochrome and stainless steel production using models and modelling as a connection between the phenomena and the applications. The role of the modelling - as well as other methods of research and development - as a connective link between the applications and phenomena is illustrated in Figure 1. It should be noted that due to sake of clarity Figure 1 does not contain all the applications, methods nor phenomena that could be related to the production of ferrochrome and stainless steels. Its purpose is to be merely an example of how the link between the applications and phenomena is created via modelling, analyses and experiments.

Although one aim for this chapter has been to present the current state of the modelling concerning the refining processes in the production of ferrochrome and stainless steel, the main goal is not to give a comprehensive outlook on all the models that are being used in the modelling of the stainless steel production processes, but to illustrate the methods,

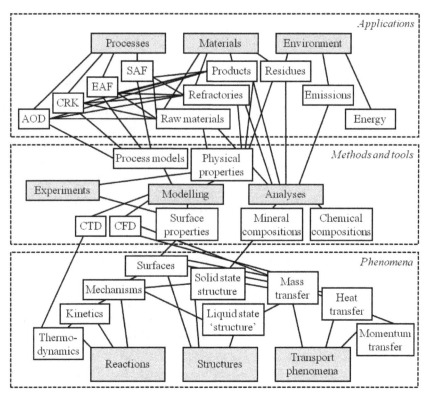

Fig. 1. Role of R&D methods as a connective link between the applications and the phenomena.

variables and dependences that are required to know in the optimization of these processes. In this sense some very useful process models may have been omitted here if it is considered that the questions mentioned above have already been answered by using other models as examples. All the examples presented in this chapter have been published before either by the authors and their colleagues or by other scientists in scientific papers. Although most of the examples presented in this chapter are related to the processes used in the Outokumpu Chrome and Outokumpu Stainless Tornio works, the results can be considered as general due to phenomenon-based nature of the models.

2. Stainless steel and ferrochrome

The amount of stainless steel that is annually produced worldwide has increased from approximately 19 million metric tons in 2001 to approximately 31 million metric tons in 2010 (International Stainless Steel Forum [ISSF], 2011a). Although this is not much when compared to the total annual production of crude steel (being approximately 1414 million metric tons in 2010; World Steel Association, 2011), there are certain areas of application in which the stainless steels have an essential role due to their resistance against corrosion (e.g. in construction, transport, process equipment and catering appliances). Corrosion resistance is achieved by using chromium as an alloying element. In oxidizing conditions a thin yet dense layer is formed on the surface of the metal if the chromium content of the metal is over approximately 12 w-% (e.g. Wranglén, 1985). This layer passivates the metal and prevents corrosion in several corrosive media.

In addition to chromium - whose content may vary from 12 w-% to nearly 30 w-% - stainless steels also contain other alloying elements that are used to adjust the structure as well as physical and chemical properties of the final product. The most commonly used grades of stainless steels are austenitic steels in which nickel is a third major component in addition to iron and chromium. Other major stainless steel categories are ferritic stainless steels (no nickel), martensitic stainless steels (less chromium in comparison to e.g. austenitic grades) and duplex steels (more chromium, some nickel). Additionally, some other elements are used in certain grades to enhance certain critical properties (e.g. molybdenum to enhance the corrosion resistance against acids) or to stabilize a certain structure (e.g. manganese to replace nickel as austenite stabilizer). Typical contents of main elements for different stainless steel grades are presented in Table 1.

Grades	Cr	Ni	C	Si	Mn	Others	AISI
Austenitic	16-20	> 8	< 0.1	< 1	< 2	Mo,Nb, Ti	e.g. 304, 316,317
Ferritic	11.5-17	-	< 0.1	< 1	< 1	Al,Ti	e.g. 405
Martensitic	11.5-17	- / < 2.5	0.1-0.4	< 1	< 1		e.g. 403, 410,414
Duplex	> 22	4-6				Mo	
Cr-Mn Austenitic	16-18	1-8	< 0.1	< 1	5-10		

Table 1. Typical compositions (in w-%) of the different kind of stainless steel grades (collected from various sources).

Since all stainless steels contain chromium, it is necessary either to alloy chromium or to use chromium-containing raw materials (e.g. ferrochrome and stainless steel scrap) in the production of stainless steels. From the economic point of view, the latter option is more suitable than the former. Concerning these raw materials mentioned above, ferrochrome is usually produced from chromite ores and concentrates, in which both of the stainless steel's main components - i.e. iron and chromium - are present. Since the occurence - and hence also the availability - of these raw materials is globally rather limited (Riekkola-Vanhanen, 1999), the use of internal and/or recycled merchant stainless steel scrap as a raw material is essential in the production of stainless steels. For instance, in 2010 over 5 million metric tons of stainless steel scrap were exported worldwide (ISSF, 2011b) and it is estimated that the average content of recycled material in new stainless steel is over 60 % (ISSF, 2011c). The end-of-life recycling rate of stainless steels is approximately 70 % (Reck et al., 2010).

Typical compositions of ferrochrome as well as chromite ore and concentrates are presented in Table 2. The compositions of the stainless steel scrap correspond with the compositions presented already in Table 1.

	Cr	Fe	C	Si	O	Others
Ferrochrome	53-55	33 / 37	7	3-5	-	
Chromite ores and concentrates	25-31	12-19	-	1-6	32-37	Al 4-15 Mg 7-13

Table 2. Typical compositions (in w-%) of ferrochrome as well as chromite ores and concentrates (collected from various sources).

3. Production of ferrochrome and stainless steel

As it was mentioned above, stainless steels are solutions of iron and chromium as well as other alloying elements that are used in order to obtain wanted chemical and mechanical properties for the final products. In the production of these highly alloyed steels it is important to utilize processes in which the yields of the valuable elements - such as chromium - are as high as possible. For the unit operations in which undesirable elements - such as carbon, silicon or sulphur - are to be removed, this obviously sets certain challenges that are not encountered in the production of low-alloyed steels. Taking decarburization as an example, it is necessary not only to decrease the amount of carbon to a required level, but to prevent excessive oxidation of chromium at the same time. This means that although there are certain similarities in the production routes of stainless and low-alloyed steels, some unit operations - such as decarburization mentioned above - are usually executed using different kind of process solutions in comparison to low-alloyed steelmaking.

One of the key issues in the production of stainless steels is in fact to solve how to execute this decarburization without the simultaneous oxidation of chromium; i.e. how to promote the oxidation reaction of carbon presented in Equation (1) and to prevent the oxidation reaction of chromium presented in Equation (2):

$$\underline{C}_{Fe} + \underline{O}_{Fe} = CO \text{ (g)} \tag{1}$$

$$2\,\underline{Cr}_{Fr} + 3\,\underline{O}_{Fe} = (Cr_2O_3)_{Slag} \tag{2}$$

Equilibrium constants (K) for reactions (1) and (2) are presented in Equations (3) and (4), respectively:

$$K_{(1)} = \frac{p_{CO}}{a_C \cdot a_O} = e^{-\frac{\Delta G^0_{(1)}}{R \cdot T}} \tag{3}$$

$$K_{(2)} = \frac{a_{Cr_2O_3}}{a_{Cr}^2 \cdot a_O^3} = e^{-\frac{\Delta G^0_{(2)}}{R \cdot T}} \tag{4}$$

in which p_i represents the partial pressure of component i, a_i representes the activity of component i, $\Delta G^0_{(x)}$ is standard Gibbs free energy for reaction (x) [in J·mol⁻¹ or cal·mol⁻¹], R is the gas constant [in J·mol⁻¹·K⁻¹ or cal·mol⁻¹·K⁻¹] and T is temperature [in K].

It is seen from Equation (3) that in order to enhance the oxidation of carbon (at constant temperature) one has to either increase the activity (and amount) of oxygen or to decrease the partial pressure of carbon monoxide in the system. According to Equation (4) the increase in the activity of oxygen enhances also the oxidation of chromium and hence the only option is to decrease the partial pressure of CO. In order to do this, it is possible either to dilute the gas by blowing inert gases such as nitrogen and/or argon into the system or to decrease the total pressure of the system. The former method is used in AOD converters (Argon Oxygen Decarburization), whereas the latter is utilized in VOD converters (Vacuum Oxygen Decarburization).

In order to properly execute the decarburization, either AOD or VOD converter is used in nearly all stainless steel production routes. The AOD converter is more commonly used as approximately three quarters of the stainless steel world production is manufactured using an AOD. (Choulet & Masterson, 1993; Jones, 2001) On the other hand, the charge sizes of the VOD converter are usually smaller than the ones of the AOD, and therefore it is often used in the steel works in which the annual production is smaller.

Evidently stainless steel production consists of several other unit operations in addition to AOD (or VOD). These other processes are chosen based on the steel grades that are being produced as well as the raw materials that are being used. In the scrap-based stainless steelmaking, electric arc furnace (EAF) is used to melt the scrap, after which the molten metal is further processed in AOD, VOD and/or various ladle treatments depending on the requirements set by the product. In other words the melting of the material and the metallurgical processing (*i.e.* mainly decarburization) are separated from one another into two independent processes. If ferrochrome is used as a raw material, it is also melted in the EAF, unless the production of ferrochrome is located in the same area, which enables the transfer of molten material from the ferrochrome plant to the stainless steel plant. As for ferrochrome, it is most commonly produced in submerged arc furnaces (SAF) using chromite ores and/or concentrates as raw materials. A so-called (ferro)chrome converter (CRK) can also be used as a pretreatment process between SAF and AOD, if ferrochrome is transfered to stainless steel plant in a molten state.

A schematic presentation of the AOD process is presented in Figure 2. Due to geometrical similarities between the AOD and CRK processes, Figure 2 also gives an impression of what a CRK process looks like.

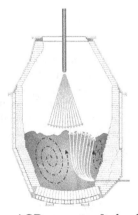

Fig. 2. Schematic presentation of an AOD converter. It should be noted that the ferrochrome converter is geometrically very similar to AOD presented in this figure.

The AOD (or VOD) is followed by e.g. ladle treatments, continuous casting, slab grinding, hot and cold rolling as well as pickling and annealing. However, these processes are omitted in this study, since the focus is on the refining processes preceeding these operations.

An example of the stainless steel production route is illustrated in Figure 3 in which the process flow sheet of the Outokumpu Stainless and Outokumpu Chrome Tornio Works in Finland is presented. The reason for choosing Tornio Works as an example is not only the fact that most of the models presented in the following section are validated using data from Tornio, but also because as an only integrated stainless steel plant in Europe (Riekkola-Vanhanen, 1999) it features most of the unit operations that are relevant for stainless steelmaking. In fact, VOD is the only significant process of stainless steelmaking that is not included in the Tornio works process route.

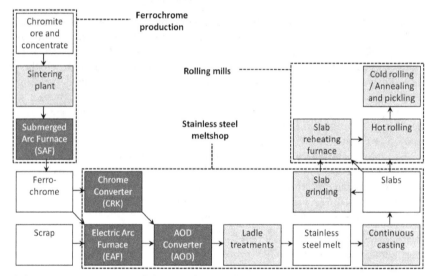

Fig. 3. Schematic presentation of the stainless steel production at the Outokumpu Chrome and Outokumpu Stainless Tornio Works (materials are presented in white boxes and processes in grey boxes).

The following section focuses on the refining processes of the ferrochrome and stainless steel production illustrated as dark grey boxes in Figure 3 (*i.e.* SAF, EAF, CRK and especially AOD), whereas the process steps illustrated as light grey boxes are omitted. From the modelling's point of view, the VOD converter, which is not included in Figure 3 and which is omitted in the next section, has many similarities with the AOD and therefore many things concerning the modelling of AOD can be applied to VOD, too, as long as the effect of reduced pressure can be taken into account.

4. Modelling of the refining processes

As seen already in Figure 1, the target of the modelling may be either processes (i.e. process modelling) or material properties and phenomena (i.e. phenomenon-based modelling). The phenomenon-based modelling may be further divided into analytical, numerical and physical modelling. Analytical models consist of pure equations describing the effect of certain variables on the modelled property in a way that is solvable analytically. In numerical models these interdependencies between the properties and certain variables are to be solved using numerical optimization and minimisation methods. Typical examples of numerical modelling are computational fluid dynamics (CFD) and computational thermodynamics (CTD). In addition to these computational models, some models are based on physical analogies between a real system and a physical model. An example of this kind of physical modelling is the modelling of metal and/or slag flows using small scale water models.

The purpose of this section is to present the essential features that need to be considered in the modelling of the SAF, EAF, CRK and AOD processes used in the production of ferrochromium and stainless steels (cf. Table 3). Although few more comprehensive models

Process	SAF	EAF	CRK	AOD
Main purpose	Reduction of chromite	Melting of scrap	Silicon removal	Decarburization
Key issues to be modelled	Reduction of chromite into ferrochrome Energy consumption	Prevention of oxidation of chromium Energy consumption Slag foaming Removal of carbon	Removal of silicon (and partly carbon) from metal Prevention of oxidation of chromium	Decarburization Control of nitrogen pickup Prevention of oxidation of chromium Sulphur removal
Influencing factors	Reduction reactions Electric potential Chemical and mineralogical composition and structure of raw materials	Heat transfer Electric potential Formation, stability and behaviour of electric arc Viscosities	Oxidation and reduction reactions (metal, slag) Reaction surface and surface properties Convection Viscosities	
Variables needed to be considered in the modelling	Amounts, compositions and grain size distributions of raw materials	Amounts and compositions of raw materials FeSi and coke additions Electric currents	Blowing practice (durations of different stages; proportions of oxygen, nitrogen and argon in different stages; lance height; top/bottom blowing) Lime and scrap additions	

Table 3. Essential features in the modelling of the phenomena taking place in the production processes of ferrochrome and stainless steel.

are also presented in addition to phenomenon-based modelling, the goal is not to present any factory-specific process models, but to consider the phenomena - as well as the factors influencing these phenomena - that need to be understood in the control of these processes. The modelling as well as its results and validation concerning each individual process is presented in more detail in subsections 4.1 to 4.4.

4.1 Submerged arc furnace, SAF

The purpose of SAF is to produce ferrochrome by reducing chromite ores and concentrates. Whereas iron oxides can be reduced into iron with carbon in for example blast furnaces, the reduction of chromium with carbon requires temperatures too high to be implemented in blast furnace type ovens. SAF, which is the best available technology for high carbon ferrochrome production, uses electric current to increase temperature and to enhance the reduction reactions of the chromite. (Riekkola-Vanhanen, 1999) Concerning the modelling of this process, it is necessary to know what is the influence of the used electric current on the physical and chemical properties of the used materials (i.e. ores, concentrates, coke, silica) as well as on the physical and chemical phenomena taking place in the furnace (i.e. chemical reactions, heat transfer, flow patterns, electric conduction, and so on). With this information it is possible to estimate, how the amount and quality of the product as well as the energy consumption of the process are influenced if the electric current or the amounts, compositions and/or grain size distributions of the raw materials were being changed.

Although the reduction reactions of the chromite have been studied by e.g. Dawson & Edwards (1986), Niayesh & Dippenaar (1992), Xiao et al. (2004), Chakraborty et al. (2005) and Zhao & Hayes (2010), the influece of the electric current on the reduction reactions is usually omitted as anything else but a heat source (Rousu et al., 2010). Experiments conducted by Rousu et al. (2010) indicated the enhancing effect of electric current on the reduction reactions thus giving a little information concerning the dependancies between the current and the reactions. On the other hand, based on the analyses of the samples taken from the actual SAF process, Ollila et al. (2010) have concluded that the electrical properties of the materials change drastically as the reduction of the chromite proceeds. The same study also indicated the circulation phenomena of zinc, sulphur and the alkali that need to be considered in the modelling of the SAF process. (Ollila et al., 2010) However, the amount of information at the moment is not sufficient enough for proper modelling of the SAF phenomena and more experimental research is required before comprehensive models concerning the SAF could be presented. An extensive review on the research concerning the chemical reactions inside the SAF in the ferrochrome production is provided by Hayes (2004).

4.1.1 SAF in the production of other ferroalloys

At this point it is worth mentioning that the SAF process is used in the production of other ferroalloys, too. For example ferrosilicon and ferromanganese are being produced with SAF type processes. (e.g. Batra, 2003; Slizovskiy & Tangstad, 2010; Tang & Olsen, 2004) Concerning the modelling of these processes, it is necessary - as it is in the production of ferrochrome - to know the influence of the electric current on the physical and chemical properties and phenomena within the furnace. In the thermodynamic modelling of the ferroalloys production systems, the most important phases to be modelled are the liquid slag and the metal alloy itself (Tang & Olsen, 2004). Despite the existing similarities, production of each ferroalloy has also its own characteristic features that needs to be taken into account in the modelling of these processes.

The production of ferrosilicon consists of quartzite reduction using charcoal, pitch coke, petroleum coke, coke breeze or anthracite as reducing agents. The iron is provided to the system with the addition of steel chips, iron ores or mill scales. The obvious difference to the ferrochrome production is the higher stability of SiO_2 in comparison to Cr_2O_3 or $FeCr_2O_4$. This means that more energy is required for the reduction of SiO_2. In addition to this, one has to consider the possible formations of SiC and gaseous SiO. Thermodynamic calculations concerning the reduction of quartzite with carbon indicate that the formation of SiC may become a problem when the silicon content of the metal exceeds 22 w-%. On the other hand, the formation of SiO-gas my decrease the silicon yield in higher temperatures. (Batra, 2003)

In the modelling of the ferromanganese production, one also has to consider the influences of potassium and zinc compounds that are always present in the raw materials. The concentrations of these elements may increase in the furnace due to circulation phenomenon and hence lead to problems such as coke particle disintegration, formation of hard bank materials and irregular flows in the furnace. (Slizovskiy & Tangstad, 2010)

Concerning the modelling of the chemical and physical properties of the slags in the ferroalloys production, a comprehensive reviews on the topic has been provided by Lehmann et al. (2004) and Jahanshahi et al. (2004).

4.2 Electric arc furnace, EAF

The main purpose of EAF is to melt scrap with electricity and hence make it possible to perform melting and decarburization in two separate unit operations (i.e. EAF and AOD), although it is possible to perform decarburization also in the EAF. Since melting is in question, one of the most important things to be considered is obviously the heat transfer between the electric arc and the metal. Other key elements in the control of the EAF process are to prevent the oxidation of chromium (if/when chromium-alloyed scrap is being used), to stabilize the arc and to minimise specific energy consumption as well as to create a foamy slag that protects the refractory materials from the electric arc, lowers energy consumption, stabilizes the arc and insulates the metal from the atmosphere. (Alexis et al. 2000; Arh & Tehovnik, 2007; Kerr & Fruehan, 2000) The process variables that can be used to control these phenomena are the amounts and compositions of the raw materials, FeSi and coke additions, use of slag formers and the use of electricity at different stages of the process.

Heat transfer between the arc and the metal with different currents and arc lengths has been studied with computational fluid dynamics (CFD) in which the arc is treated as a fluid with temperature-dependent properties. The model in which various mechanisms of heat transfer (i.e. radiation, convection, condensation and energy tranported by electrons) were taken into account indicated that although the contribution of electrons on the heat transfer is significant at the center of the system, the radiation and convection seem to be more dominant in general. The longer the arc, the more dominant is the role of radiation in the system. (Alexis et al. 2000) In addition to CFD, physical modelling has also been used in the investigation of the injection treatments in the EAF. The effect of lance height, lance angle and gas flow on the cavity that is formed at the melt surface when blowing oxygen into EAF is one of the applications for which the physical water models has been used in the context of EAFs. (Whitney 2003)

The oxidation of chromium in the EAF process has been studied with thermodynamic calculations, according to which the oxidation of chromium may be decreased with aluminium and silicon additions, since these two elements have a higher affinity for oxygen than chromium. On the other hand, the oxidized chromium may be reduced from the slag back to metal by using ferrosilicon, carbon, calcium carbide or aluminium. (Arh & Tehovnik, 2007; Park et al., 2004) When modelling the oxidation reactions, one always has to consider the competing oxidation reactions of carbon, chromium and silicon as well as in some cases aluminium and titanium.

Studies concerning the reduction of Cr_2O_3 as well as other oxides (MnO, FeO and SiO_2) from the EAF slags by using either carbon or aluminium as reductant indicate that the reduction is controlled by diffusional mass transfer in the slag phase. The mass transfer coefficient describing the diffusion rate of a certain oxide in the slag was found to be largest for MnO and clearly smallest for SiO_2. The coefficients describing the mass transfers of Cr_2O_3 and FeO are close to one another and slightly smaller than the one for MnO, although considerably higher than the one for SiO_2. However, the overall reduction degrees of these oxides decresed in the order of MnO (90 %), Cr_2O_3 (70 %), SiO_2 (59 %) and FeO (40 %). (Park et al., 2004)

EAF slag's ability to generate foam from injected or generated gas can be quantified with the foam index, Σ, for which the mathematical definition as well as the correlation with the slag properties is given in equation (5) (Ito & Fruehan, 1989; Kerr & Fruehan, 2000, 2002, 2004; Zhang & Fruehan, 1995):

$$\Sigma = \frac{H_f}{V_g^s} = \frac{H_f}{\dfrac{Q}{A}} = \frac{H_f \cdot A}{Q} = 115 \cdot \frac{\mu^{1.2}}{\gamma^{0.2} \cdot \rho \cdot D^{0.9}} \tag{5}$$

in which H_f indicates the foam height [in m] and V_g^s is the superficial gas velocity [in $m \cdot s^{-1}$] defined as a ratio between the gas flow rate, Q [in $Nm^3 \cdot s^{-1}$], and the cross sectional area of the vessel or crucible, A [in m^2]. In the latter part of the equation (5), μ represents slag's bulk viscosity [in Po], γ slag's surface tension [in $N \cdot m^{-1}$], ρ slag's density [in $kg \cdot m^{-3}$] and D the bubble diameter [in m]. (Ito & Fruehan, 1989; Kerr & Fruehan, 2000, 2002, 2004; Zhang & Fruehan. 1995).

It has been observed that as the Cr_2O_3 content of the slag is increased, the foam index stays constant for a while (up to approximately 8 %), after which there is a small increase in the values of the foam index in the Cr_2O_3 content range of approximately 8 to 13 %. If the Cr_2O_3 content is further increased, the excessive Cr_2O_3 begins to precipitate as a solid phase which increases the viscosity of the slag as seen from equation (6) and hence decreases the values of the foam index very rapidly. The critical Cr_2O_3 content over which the foam index decreases depends on temperature, slag composition and oxygen potential. (Arh & Tehovnik, 2007; Kerr & Fruehan, 2000, 2002, 2004)

$$\mu = \mu_0 \cdot \left(1 + 5.5 \cdot \xi\right) \tag{6}$$

in which μ is the bulk viscosity [in Po], μ_0 is the viscosity of the pure liquid [in Po] and ξ is the volume fraction of solid particles (Kerr & Fruehan, 2000, 2002). It has been concluded that the occurence of solid Cr_2O_3 particles in the slag is one of the reasons why slag foaming in stainless steelmaking is not as efficient as in the production of low-alloyed steels (Arh & Tehovnik, 2007; Kerr & Fruehan, 2000, 2002, 2004).

Another difference concerning the slag foaming between the productions of stainless and low-alloyed steels is the amount of FeO in the slag. In the stainless steelmaking the FeO content is usually considerably lower and hence the CO gas which is required for the slag foaming is not generated via the reaction between FeO and carbon, but via reactions of other oxides (e.g. CrO and Cr_2O_3) and carbon. Since the reaction rates for the reactions between chromium oxides and carbon are significantly lower than the one for reaction between FeO and carbon, it can be concluded that the CO formation is considerably slower when considering the production of stainless steels. (Arh & Tehovnik, 2007; Kerr & Fruehan, 2000, 2002, 2004)

The modelling of the electric arc itself and its influence on the chemical and physical phenomena occuring in the EAF is very complicated (cf. SAF). Even if the nature of the arc and its influence on e.g. chemical reactions were known, the modelling would be little more than educated estimations due to extremely high temperatures (which cannot be measured accurately) and the lack of material data and information in such high temperatures.

4.3 Ferrochrome converter, CRK

Ferrochrome converter is a link between the production of ferrochrome and stainless steel production. Its purpose is to enable the transfer of ferrochrome in a molten state from the ferrochrome plant to the stainless steel melting shop and to enable the utilisation of ferrochrome's chemical energy for scrap melting. During the CRK process oxygen is blown into the metal, due to which the silicon content of the metal is decreased from 4 to 5 w-% to less than 0.5 w-%. At the same time, the carbon content is decreased from approximately 7 w-% to approximately 3 w-%. Further decarburization would lead into excessive oxidation of chromium. Exothermic oxidation reactions of silicon and carbon release energy that is used to heat the metal and to melt scrap and slag formers. Although the purpose of CRK is different to more commonly used AOD converter, these two converters are geometrically very similar to each other with almost identical shape and both having sidewall tuyeres in addition to the top lance. (Fabritius & Kupari, 1999; Fabritius et al., 2001a, 2001b; Heikkinen et al., 2010b, 2011; Virtanen et al. 2004)

4.3.1 Thermodynamics

In the modelling of the CRK process it is essential to consider how the oxidation reactions of silicon, carbon and chromium (i.e. removal of silicon and carbon as well as prevention of chromium oxidation) are influenced by process variables such as metal and slag compositions as well as fluid flows and properties (e.g. viscosities). Knowledge concerning the interdependencies between the above mentioned oxidation reactions and process variables enables the optimization of process parameters such as blowing practices and lime and scrap additions.

Thermodynamic computations validated with process samples from the Outokumpu Stainless Tornio steelworks have been used to estimate the chemical driving forces for the oxidation reactions of silicon, carbon and chromium (cf. equation (7) in which X represents Si, C and Cr) at the different stages of the process. (Heikkinen et al., 2010b, 2011)

$$\frac{2a}{b} \underline{X}_{Fe} + O_2(g) = \frac{2}{b} \begin{cases} (X_aO_b) \\ X_aO_b(g) \end{cases}$$ (7)

Thermodynamic equilibria of these oxidation reactions were compared with each other by using partial pressures of oxygen (p_{O2}) as parameters describing the equilibria. Smaller value of p_{O2} for a certain oxidation reaction indicate that less oxygen is needed for oxidation when compared to a reaction for which the value of p_{O2} is larger. The partial pressures were calculated using equations (8) and (8') the former being the equation for silicon and chromium (for which the oxidation products end up in slag) and the latter being the equation for carbon (for which the oxidation product is a gas). (Heikkinen et al., 2010b, 2011)

$$p_{O_2} = \frac{a_{X_aO_b}^{2/b}}{f_X^{2a/b} \cdot X_X^{2a/b}} \cdot e^{\left(\frac{\Delta G^0_{(7)}}{R \cdot T}\right)}$$ (8)

$$p_{O_2} = \frac{p_{X_aO_b}^{2/b}}{f_X^{2a/b} \cdot X_X^{2a/b}} \cdot e^{\left(\frac{\Delta G^0_{(7)}}{R \cdot T}\right)}$$ (8')

in which p_i represents the partial pressure of component i, a_i representes the activity of component i, f_i representes the activity coefficient of component i, X_i representes the mole fraction of component i, $\Delta G^0_{(x)}$ is standard Gibbs free energy for reaction (x) [in J·mol^{-1} or cal·mol^{-1}], R is the gas constant [in J·mol^{-1}·K^{-1} or cal·mol^{-1}·K^{-1}] and T is temperature [in K]. The compositions needed for the calculations were obtained from the analyses carried out for the process samples from Tornio works. Thermodynamic values needed for the calculations were calculated by using quadratic formalism by Ban-Ya (1993) and Xiao et al. (2002) for the slag phase and unified interaction parameter formalism by Pelton & Bale (1986) for the metal phase. Some values were also obtained from the databases of thermodynamic software such as FactSage and HSC Chemistry. (Heikkinen et al., 2010b, 2011)

The results of the modelling were used to estimate the oxidation order of silicon, carbon and chromium in different process conditions. The results indicated that with the high silicon contents in the early stages of the CRK process silicon is most likely to be oxidized before carbon and chromium. As the silicon content decreases below approximately 0.3-0.6 w-%, the oxidation of carbon becomes more favourable and the carbon content of the metal begins to decrease. If the oxygen blowing is continued even further, chromium begins to oxidise excessively if the carbon content is decreased below approximately 2.5 w-%. This means that the reduction of Cr_2O_3 from the slag is required in order to produce low carbon ferrochrome with the CRK process. (Heikkinen et al., 2010b, 2011)

4.3.2 Flow patterns

Obviously a comprehensive look on the phenomena taking place in the CRK cannot be given by a thermodynamic study alone and therefore it is also necessary to consider the influence of blowing parameters on the flowing patterns of the liquid metal. This is important not only due to flow pattern's influence on the reactions, but also due to optimization of sufficient mixing and minimisation of unwanted spitting and splashing. The influence of the lance height and amount of holes in the lance as well as the amounts of gas blown through the lance and through the tuyeres at the sidewall of the converter have been studied using physical water models built to correspond the actual CRK process in Tornio, which was also used to validate the results of the modelling. (Fabritius & Kupari, 1999; Fabritius et al., 2001a, 2001b; Virtanen et al. 2004)

The use of physical water models to investigate the flow patterns of a converter is based on an assumption according to which the lack of information concerning the extremely turbulent conditions inside the converter limit the numerical modelling of the process. Therefore it is considered to be more meaningful to construct physical models with dynamic similarities to real processes. This way it is not necessary to know mathematical expressions for all the interdependencies between the process parameters and variables as long as it is possible to ensure the dynamic similarities between the model and the real process. Confirmation of these similarities is defined by using dimensioless numbers such as Froude's, Reynold's, Weber's, Ohnesorge's and Morton's numbers which describe the ratios between different forces in the model and in the real process. The values of a certain dimensioless number defined for the model and for the real process should be set as close to equal as possible by choosing the model geometry in a way that corresponds with the real process. The Froude number, which describes the ratio between gravitational and inertial forces and which has an essential role in the physical modelling of the CRK process, is presented as an example of a dimensioless number in equation (9). (Fabritius & Kupari, 1999; Fabritius et al., 2001a, 2001b; Virtanen et al. 2004)

$$Fr = \frac{\rho_g}{\rho_l} \cdot \frac{U^2}{gD_c} \tag{9}$$

in which ρ_g and ρ_l are the densities of the gas and liquid phases [in kg·m^{-3}], respectively, g is gravitational constant [in m·s^{-2}], U is the velocity of the gas [in m·s^{-1}] and D_c is the diameter of the converter [in m] (Fabritius & Kupari, 1999).

As the name suggests, water is used in physical water models to illustrate the metal due to kinematic similarities between the liquid water at room temperature and liquid steel at approximately 1600 °C. On the other hand, its transparency helps to make observations inside the process.

The results of the physical modelling have indicated that the rates of the oxidation reactions of silicon and carbon may be increased due to enhanced mixing by increasing the total amount of gas blowing and by increasing the amount of inert gas that is blown through the sidewall tuyeres. These changes have also been noticed to increase the penetration of the tuyere gas jets and hence decrease the occurrence of the so-called back-attack phenomenon that is detrimental to the refractory materials close to the tuyeres. The positive influence of the changed blowing practice has also been verified with the process experiments. Another results from the flow pattern studies suggest that the removal of silicon can be improved by using a higher lance position at the first stage of the blowing, whereas the removal of carbon in the later stages is improved with a lower lance position. The oxidation of chromium was observed to decrease with the use of a top lance with wider nozzle angles. Furthermore, the physical models have been used to illustrate how the change from one-hole lance to three-hole lance decreases unwanted spitting and splashing in the converter and hence enables the use of greater gas volumes - and increased productivity - without increasing the amount of metal losses due to splashing. (Fabritius & Kupari, 1999; Fabritius et al., 2001a, 2001b; Virtanen et al. 2004)

4.4 Argon Oxygen Decarburization converter, AOD

AOD is the process in which most of the world's stainless steel is being produced. Its purpose is to remove carbon - as well as silicon if CRK is not used - without excessive oxidation of chromium from the metal that consists of ferrochrome and molten scrap. Other important features concerning the AOD are nitrogen control and sulphur removal, latter of which is usually executed as a separate stage at the end of the processing. In order to control the process it is important to know how to influence oxidation reactions and how to control the nitrogen content with accessible process variables. The variables by which the AOD is usually being optimized are the blowing ratios between top and bottom blowing, lance positions and oxygen/nitrogen/argon -ratios in the gas during the different stages of the blowing as well as the durations of stages themselves. Furthermore, it is possible to control the process via additions of scrap, lime and alloying elements. Since the carbon is removed much more efficiently when compared to CRK, a reduction stage in which the oxidized chromium is reduced is needed after the decarburization at the end of the processing.

Due to similarities with the CRK process presented above, it is unnecessary to repeat certain things that are already covered in the previous section.

4.4.1 Thermodynamics

Thermodynamic considerations of the reaction equilibria in the AOD are more or less similar to the CRK considerations, since both of the considered systems consist of same phases (i.e. Fe-Cr-Si-C-based metals, silicate slags and gas) and the considered reactions are the same (i.e. oxidation reactions of silicon, carbon and chromium). Therefore it is safe to assume that if a certain thermodynamic model is suitable for the melts in the CRK, it is suitable for the considerations of the AOD, too. Higher metal concentrations of chromium, silicon and carbon in the CRK may restrict the application of some of the AOD models into the CRK systems, but the problem should not exist vice versa.

Obviously the consideration of the oxidation reactions is not the only application for which thermodynamic modelling can be applied in the AOD, but other applications of less relevance such as solidification of the slag and its utilization to protect the refractory material (Heikkinen et al., 2004) are omitted here.

4.4.2 Reaction rates

Although thermodynamic modelling helps us to define the chemical equilibria for the reactions taking place in the AOD, it gives us no information concerning the rates and mechanisms of these reactions. Based on the results of the studies concerning the decarburization rates of Fe-Cr-melts with a top blown oxygen, it has been concluded that the decarburization phenomenon can be classified into two separate stages, during which the decarburization rates can be modelled by using equations (10) and (11), respectively (Kitamura et al., 1986):

$$-\frac{d[\%C]}{dt} = \alpha \tag{10}$$

$$-\frac{d\log[\%C]}{dt} = \beta \tag{11}$$

in which [%C] is the carbon content of the metal [in w-%], t equals time and α and β are condition-specific constants. At the beginning of the decarburization (cf. equation (10)) the carbon content of the metal is still high and during this first stage the decarburization rate is increased with increased oxygen supply and increased bath stirring intensity. At the same time, the oxidation of chromium is also increased with increased oxygen supply, but decreased with increased stirring intensity. Below the critical carbon content the decarburization is controlled by mass transfer of carbon and the equation (10) is no longer valid. During this second stage equation (11) is used to model the decarburization rates. At this stage both decarburization and oxidation of chromium are increased with increased oxygen supply and stirring intensity. (Kitamura et al., 1986)

4.4.3 Flow patterns

Due to geometrical similarities, the modelling of the flow patterns of the AOD process is also very similar to the one of the CRK. The things about dynamic similarities and dimensioless numbers mentioned in the context of the CRK are as relevant for the modelling of the AOD as they are for the CRK, but it is unnecessary to repeat those things here. Some of the results of physical modelling concerning the AOD and CRK flow patterns are also quite similar. For example, the positive influence of the increased gas volumes through the sidewall tuyeres on the penetration of gas jets and on the duration of the refractory materials close to tuyeres has also been observed with the AOD process (Fabritius et al., 2000, 2003).

Since AOD is much more widely used in comparison to CRK, it has also been modelled much more frequently in comparison to CRK, which makes it impossible to present all the investigations here. However, some examples of the results are presented in order to give an outlook on the conlusions that have been made based on the physical modelling of the AOD process:

- The minimum gas flow that is required to prevent the occurence of the back-attack phenomenon depends on the phenomena such as melt oscillation and tuyere blockage and it could be defined by modified Froude number (Fabritius et al., 2003).

- Depending on the ratio between bath height and diameter as well as on the intensity of the gas blowing through sidewall tuyeres the main circular flow in the AOD may change its direction (Fabritius et al., 2003).
- There are two different kind of wave motions after which the melt inside the AOD converter may oscillate. The oscillation depends on sidewall blowing and according to the water models the appearance of the oscillation is most likely during the end of the decarburisation stage or during the reduction stage if tuyeres with large diameters are being used. With smaller tuyeres the penetration of the gas jets is deeper and the oscillation occurs most likely during the decarburisation stage. The modelling results concerning the oscillation have been confirmed with vibration measurements from the real AOD process in Tornio. (Fabritius et al., 2003, 2005; Kato et al., 1985; Mure et al., 2004)

In the section concerning the modelling of the flow patterns in the CRK converter it was mentioned that physical modelling is often prefered over numerical due to high complexity of the flows inside the process. Computational fluid dynamics (CFD) have nevertheless been used to simulate the flows within the AOD in some cases and the results indicate relatively good accordance with the physical modelling and real process measurements. The key issues to be considered in the CFD modelling of the AOD are model geometry, grid distribution, behaviour of free surface, introduction and removal of materials into the system, modelling of turbulence and the interaction between different phases. Additionally, one has to consider whether it is necessary to take heat transfer and chemical reactions into account. It is usually easier and faster to consider isothermal and non-reactive systems, although in some cases this might lead into oversimplification of the model. (Gittler et al., 2000; Tang et al., 2004; Tilliander et al., 2004)

4.4.4 Nitrogen and interfacial phenomena

One of the differences between the AOD and CRK processes concerns the behaviour of nitrogen. Whereas the CRK process can be considered as a pretreatment unit for the stainless steel production, in which the control of nitrogen is not yet essential, it is important to control the amount of nitrogen of the metal during the AOD process in order to be able to produce stainless steel grades with strict composition limits. In the early stages of the AOD process the blowing gas consists of pure oxygen, but in order to decrease the oxidation of chromium - and to be able to reduce Cr_2O_3 during the reduction stage - the oxygen needs to replaced by nitrogen and argon. The use of cheaper nitrogen can be justified with economic reasons in the earlier stages, but due to nitrogen pick-up (of up to 1000 ppm) it is necessary to use more expensive argon during the last stage of decarburisation and the reduction stage. In order to control the nitrogen content of the metal, it is necessary to know what is the effect of process variables on the nitrogen pick-up (during the stages in which nitrogen is blown into the system) and nitrogen removal (during the stages in which argon is used). (Heikkinen et al., 2007; Kärnä et al., 2008; Riipi et al., 2009)

Unlike the oxidation and reduction reactions which can be modelled using computational thermodynamics with tolerable accuracy, a thermodynamic model, that describes the effects of metal compositions and partial pressures of nitrogen in the gas bubbles on the nitrogen solubility of the stainless steel melts, is not sufficient for the modelling of nitrogen. Although it is important to recognize chromium's increasing effect and nickel's decreasing effect on the nitrogen solubility, it is necessary also to model the mass and heat transfer in the gas-metal systems. Since the mechanisms of both nitrogen absorption (i.e. nitrogen pick-up) and desorption (i.e. nitrogen removal) can occur in either single-site or dual-site mechanism depending on the amount of surface active elements in the metal, it is necessary to model the influences of these surface active elements such as oxygen and sulphur on the interfacial properties, nitrogen absorption and desorption rates and furthermore on the nitrogen content of the metal. (Heikkinen et al., 2007; Järvinen et al., 2009; Kärnä et al., 2008; Riipi & Fabritius, 2007; Riipi et al., 2009)

Behaviour of nitrogen is not the only thing in which the interfacial phenomena have an important role. In a systems of three separate phases (metal, slag and gas; excluding the refractory materials) the interfacial properties - of which interfacial tensions are the most essential - always have an effect on all heterogeneous phenomena. In order to describe these effects, variables such as film coefficient, θ_m, and flotation coefficient, Δ_m, have been introduced. Based on the values of these coefficients, that are calculated using equations (12) and (13), it is possible to estimate, whether the gas bubbles and metal droplets exist separately inside the slag or whether the bubbles are surrounded by metal films or metal droplets attached to the bubbles. (Fabritius et al., 2010; Minto & Davenport, 1972)

$$\theta_m = \sigma_{slag-gas} - \sigma_{metal-gas} - \sigma_{metal-slag} \tag{12}$$

$$\Delta_m = \sigma_{slag-gas} - \sigma_{metal-gas} + \sigma_{metal-slag} \tag{13}$$

in which $\sigma_{slag-gas}$ and $\sigma_{metal-gas}$ are the surface tensions of slag and metal [in $N \cdot m^{-1}$], respectively, whereas $\sigma_{metal-slag}$ is the interfacial tension between metal and slag [in $N \cdot m^{-1}$] (Fabritius et al., 2010; Minto & Davenport, 1972).

As a conclusion one could state, that with sufficient knowledge concerning thermodynamics, reaction rate kinetics as well as surface and transport phenomena, it is possible to create submodels describing these phenomena and then include them in more comprehensive process models in which for example CFD is used. (Järvinen et al., 2009; Kärnä et al., 2008)

4.4.5 Process modelling

As opposed to phenomenon-based models, in which each phenomenon is considered separately, process models aim for more holistic approach by combining several submodels into bigger entities. Because the combination of several very detailed submodels creates slow and non-robust models that usually are not very useful, the accuracy of describing an individual phenomenon with a larger process model is usually not as high as it is in the phenomenon-based models. On the other hand the process models are usually more adaptable to real processes and in some cases they might even be part of the control and automation systems. A typical feature for many process models is, that in order to be able to model the relevant phenomena by using appropriate submodels, it is necessary to define certain parameters which are required for the modelling but are not measurable or otherwise known. These kind of parameters are often used to fit the model to better correspond with the real process. For instance, one might need to know what is the bubble size (and bubble size distribution) in order to calculate the mass transfer between the molten metal and the gas phase. However, the bubble size distribution inside the converter is difficult to measure and therefore the modeller needs to be satisfied with estimated values. By trying different bubble sizes and bubble size distributions, it is possible to adjust the model in order to make it more equivalent with the reality (Järvinen et al., 2011).

Process models may be constructed in various ways based on their purpose and usually the phenomena that are considered to be most critical on the process optimization are modelled in greatest detail (e.g. chemical equilibrium, reaction kinetics, mass transfer, fluid flows, heat transfer, interfacial phenomena, and so on). This means that separate researchers may end up with very different kind of models based on the target of their modelling as well as on their own impressions. In addition to what is included in the process models and in which extent and accuracy, the models may also differ from one another based on the connections that link the submodels with each other. Process models for the AOD converter have been proposed by e.g. Fruehan (1976), Sjöberg (1994), Wei & Zhu (2002), Zhu et al. (2007), Wei et al. (2011) and Järvinen et al. (2011).

5. Conclusion

Phenomenon-based models illustrating both physical and chemical phenomena in the refining processes of ferrochrome and stainless steel have been presented with an attempt to use models as connecting links between the phenomena and the applications relevant for the production engineer. The results obtained by using these models are not only helping us to control and optimize the metal production processes, but also helping us to understand how various process parameters and variables are linked to the process outcomes via the phenomena for which the variables have an influence on.

As accurate modelling often requires time resources that are not available for engineers working daily with the processes in the steel plants, this usually has to be made by researchers working at the universities and research institutes. However, the results, no matter how accurately modelled, can only be as good as the assumptions behind them. Therefore it has been - and always will be - necessary to validate the results obtained from the models with the real process data. This means that reliable modelling is only rarely achieved without an open and long-term co-operation between the modellers and the production engineers.

Although most of the models presented in this chapter focus on individual phenemona rather than total processes, it has not prevented the use of these models as well as the results obtained from the models as parts for more comprehensive process models. In fact, many of the results concerning the CRK and AOD processes have been implemented in the control systems of these processes at the Outokumpu Stainless Tornio steelworks.

6. Acknowledgment

The authors wish to thank the personnel of the Outokumpu Chrome and Outokumpu Stainless Tornio works for long and fruitful co-operation in the modelling of the ferrochrome and stainless steelmaking processes. Special thanks are appointed to Dr. Paavo Hooli, Mr. Topi Ikäheimonen, Mr. Pentti Kupari, Mr. Pekka Niemelä, Mr. Janne Ollila, Mr. Juha Roininen and Mr. Veikko Juntunen.

The contribution of our current and former colleagues Ms. Anne Heikkilä, Mr. Lauri Hekkala, Prof. Jouko Härkki, Mr. Tommi Kokkonen, Mr. Jari Kurikkala, Mr. Petri Kurkinen, Mr. Aki Kärnä, Mr. Olli Mattila, Mr. Riku Mattila, Mr. Petri Mure, Ms. Jaana Riipi, Mr. Arto Rousu, Mr. Jari Savolainen, Mr. Pekka Tanskanen, Mr. Esa Virtanen and Mr. Ville-Valtteri Visuri is also acknowledged with gratitude.

7. References

Alexis, J.; Ramirez, M.; Trapaga, G. & Jönsson, P. (2000) Modeling of a DC Electric Arc Furnace - Heat Transfer from the Arc. *ISIJ International*, Vol. 40, No. 11, (November 2000), pp. 1089-1097.

Arh, B. & Tehovnik, F. (2007) The oxidation and reduction of chromium during the elaboration of stainless steels in an electric arc furnace. *Materials and Technology*, Vol. 41, No. 5, (September 2007), pp. 203-211.

Ban-Ya, S. (1993) Mathematical Expression of Slag-Metal Reactions in Steelmaking Process by Quadratic Formalism Based on the Regular Solution Model. *ISIJ International*, Vol. 33, No. 1, (January 1993), pp. 2-11.

Batra, N.K. (2003) Modelling of ferrosilicon smelting in submerged arc furnace. *Ironmaking and Steelmaking*, Vol. 30, No, 5, (October 2003), pp. 399-404.

Chakraborty, D.; Ranganathan, S. & Sinha, S.N. (2005) Investigations on the carbothermic reduction of chromite ores. *Metallurgical and Materials Transactions B*, Vol. 35B, No. 4, (August 2005), pp. 437-444.

Choulet, R.J. & Masterson, I.F. (1993) Secondary steelmaking in stainless steel refining. *Iron and Steelmaker*, Vol. 20, No. 6, (May 1993), pp. 45-54, ISSN 0275-8687.

Dawson, N.F. & Edwards, R.I. (1986). Factors affecting the reduction of chromite, *Proceedings of INFACON IV 4th International Ferro Alloys Congress*, pp. 105-113, Rio de Janeiro, Brazil, August 31-September 3, 1986.

Fabritius, T. & Kupari, P. (1999). Optimization of Bottom Gas Blowing in Converter by Physical Modelling, *Proceedings of SCANMET I 1st International Conference on Process Development in Iron and Steelmaking, Volume 1*, pp. 391-411, Luleå, Sweden, June 7-8, 1999.

Fabritius, T. ; Vatanen, J. ; Alamäki, P. & Härkki, J. (2000) Effect of sidewall blowing to the wear of the refractory lining in AOD, *Proceedings of the 6th Japan-Nordic countries Steel Symposium*, pp. 114-121, Nagoya, Japan, November 28-29, 2000.

Fabritius, T. ; Kupari, P. & Härkki, J. (2001a) Physical modelling of a sidewall blowing converter. *Scandinavian Journal of Metallurgy*, Vol. 30, No. 2, (April 2001), pp. 57-64.

Fabritius, T. ; Mure, P. ; Kupari, P. ; Juntunen, V. & Härkki, J. (2001b) Combined blowing with three-hole lance in a sidewall blowing converter. *Steel Research*, Vol. 72, No. 7, (July 2001), pp. 237-244.

Fabritius, T. ; Mure, P. & Härkki, J. (2003) The determination of minimum and operational gas flow rates for sidewall blowing in the AOD-converter. *ISIJ International*, Vol. 43, No. 8, (August 2003), pp. 1177-1184.

Fabritius, T. ; Kurkinen, P. ; Mure, P. & Härkki, J. (2005) Vibration of argon-oxygen decarburisation vessel during gas injection. *Ironmaking and Steelmaking*, Vol. 32, No. 2, (April 2005), pp. 113-119.

Fabritius, T. ; Riipi, J. ; Järvinen, M. ; Mattila, O. ; Heikkinen, E-P. ; Kärnä, A. & Kurikkala, J. (2010) Interfacial phenomena in metal-slag-gas system during AOD process. *ISIJ International*, Vol. 50, No. 6, (June 2010), pp. 797-803.

Fruehan, R.J. (1976) Reaction Model for the AOD Process. *Ironmaking and Steelmaking*, Vol. 3, No. 3, pp. 153-158.

Gittler, P. ; Kickinger, R. ; Pirker, S. ; Fuhrmann, E. ; Lehner, J. & Steins, J. (2000) Application of computational fluid dynamics in the development and improvement of steelmaking processes. *Scandinavian Journal of Metallurgy*, Vol. 29, No. 4, (August 2000), pp. 166-176.

Hayes, P.C. (2004). Aspects of SAF smelting of ferrochrome, *Proceedings of INFACON X 10th International Ferroalloys Congress*, pp. 1-14, ISBN 0-9584663-5-1, Cape Town, South Africa, February 1-4, 2004.

Heikkinen, E-P. ; Fabritius, T. ; Kokkonen, T. & Härkki, J. (2004) An experimental and computational study on the melting behaviour of AOD and Chromium converter slags. *Steel Research International*, Vol. 75, No. 12, (December 2004), pp. 800-806.

Heikkinen, E-P. ; Riipi, J. & Fabritius, T. (2007). A computational study on the nitrogen content of liquid stainless steel in equilibrium with Ar-N_2-atmosphere, *Proceedings of 7th International Conference on Clean Steel*, pp. 436-443, Balatonfüred, Hungary, June 4-6, 2007.

Heikkinen, E-P. ; Fabritius, T. & Riipi, J. (2010a) Holistic Analysis on the Concept of Process Metallurgy and Its Applications on the Modelling of the AOD Process. *Metallurgical and Materials Transactions B*, Vol. 41B, No. 4, (August 2010), pp. 758-766.

Heikkinen, E-P.; Ikäheimonen, T. ; Mattila, O. & Fabritius, T. (2010b). A thermodynamic study on the oxidation of silicon, carbon and chromium in the ferrochrome converter, *Proceedings of INFACON XII 12th International Ferro Alloys Congress*, pp. 229-237, ISBN 978-952-92-7340-9, Helsinki, Finland, June 6-9, 2010.

Heikkinen, E-P.; Ikäheimonen, T. ; Mattila, O. ; Fabritius, T. & Visuri, V-V. (2011). Behaviour of silicon, carbon and chromium in the ferrochrome converter - a comparison between CTD and process samples, *Proceedings of EOSC 2011 6th European Oxygen Steelmaking Conference*, Stockholm, Sweden, September 7-9, 2011.

ISSF (International Stainless Steel Forum). (1.7.2011a). Stainless and Heat Resisting Steel Crude Steel Production, In: International Stainless Steel Forum, 20.9.2011, Available from:
<http://www.worldstainless.org/Statistics/Crude/2010.html>

ISSF (International Stainless Steel Forum). (20.7.2011b). Foreign Trade Flow Stainless Steel Scrap in 2010. In: International Stainless Steel Forum, 27.9.2011, Available from: <http://www.worldstainless.org/Statistics/Foreign+trade+scrap/>

ISSF (International Stainless Steel Forum). (3.10.2011c). What Makes Stainless Steel a Sustainable Material? In: International Stainless Steel Forum, 11.10.2011, Available from:
<http://www.worldstainless.org/NR/rdonlyres/B2393030-071D-48F3-A76C-6FA8B5B0D59B/6036/WhatMakesStainlessSteelaSustainableMaterial.pdf>

Ito, K. & Fruehan, R. (1989) Study on the Foaming of CaO-SiO$_2$-FeO Slags: Part I. Foaming Parameters and Experimental Results. *Metallurgical Transactions B*, Vol. 20B, No. 4, (August 1989), pp. 509-514.

Jahanshahi, S.; Sun, S. & Zhang, L. (2004). Recent developments in physico-chemical characterisation amd modelling of ferroalloy slag systems, *Proceedings of INFACON X 10th International Ferroalloys Congress*, pp. 316-332, ISBN 0-9584663-5-1, Cape Town, South Africa, February 1-4, 2004.

Jones, P.T. (2001). *Degradation mechanisms of basic refractory materials during the secondary refining of stainless steel in VOD ladles*, Katholieke Universiteit Leuven, ISBN 90-5682-297-7, Heverlee, Belgium.

Järvinen, M.; Kärnä, A. & Fabritius, T. (2009). A Detailed Single Bubble Reaction Sub-Model for AOD Process. *Steel Research International*, Vol. 80, No. 6, (June 2009), pp. 431-438.

Järvinen, M.; Pisilä, S.; Kärnä, A.; Ikäheimonen, T.; Kupari, P. & Fabritius, T. (2011) Fundamental Mathematical Model for AOD Process. Part I: Derivation of the Model. *Steel Research International*, Vol. 82, No. 6, (June 2011), pp. 650-657.

Kato, Y.; Nakanishi, K.; Nozaki, T.; Suzuki, K. & Emi, T. (1985) Wave motion of metal bath in bottom blowing converter. *ISIJ International*, Vol. 25, No. 6, (June 1985), pp. 459-466.

Kerr, J. & Fruehan, R. (2000). Foamibility of Stainless Steelmaking Slags in an EAF, *Proceedings of 58th Electric Furnace Conference and 17th Process Technology Conference*, pp. 1049-1063, Orlando, Florida, United States, November 12-15, 2000.

Kerr, J. & Fruehan, R. (2002) Foamability of Stainless Steelmaking Slags in an EAF. *Iron and Steelmaker*, Vol. 29, No. 4, (April 2002), pp. 45-52.

Kerr, J. & Fruehan, R. (2004) Additions to Generate Foam in Stainless Steelmaking. *Metallurgical and Materials Transactions B*, Vol. 35B, No. 4, (August 2004), pp. 643-650.

Kitamura, S-Y.; Okohira, K. & Tanaka, A. (1986) Influence of Bath Stirring Intensity and Top Blown Oxygen Supply Rate on Decarburization of High Chromium Molten Iron. *ISIJ International*, Vol. 26, No. 1, (January 1986), pp. 33-39.

Kärnä, A.; Hekkala, L.; Fabritius, T.; Riipi, J. & Järvinen, M. (2008). CFD model for nitrogen transfer in AOD converter, *Proceedings of SCANMET III 3rd International Conference on Process Development in Iron and Steelmaking, Volume 1*, pp. 155-161. Luleå, Sweden, June 8-11, 2008.

Lehmann, J.; Gaye, H. & Bonnet, F. (2004). Thermodynamics applied to ferro-alloys smelting, *Proceedings of INFACON X 10th International Ferroalloys Congress*, pp. 300-315, ISBN 0-9584663-5-1, Cape Town, South Africa, February 1-4, 2004.

Minto, R. & Davenport, W.G. (1972) Entrapment and flotation of matte in molten slags. *Transactions of the Institution of Mining and Metallurgy, Section C: Mineral processing & extractive metallurgy*, Vol. 81, No. 1, (March 1972), pp. C36-C42.

Mure, P.; Kurkinen, P.; Fabritius, T. & Härkki, J. (2004). Vibration Measurement of Oscillation of Melt Bath in 150-ton AOD-converter, *Proceedings of SCANMET II 2nd International Conference on Process Development in Iron and Steelmaking, Volume 2*, pp. 59-68, Luleå, Sweden, June 6-8, 2004.

Niayesh, M.J. & Dippenaar, R.J. (1992). The solid-state reduction of chromite, *Proceedings of INFACON VI 6th International Ferro Alloys Congress*, pp. 57-63, Cape Town, South Africa, March 8-11, 1992.

Ollila, J.; Niemelä, P.; Rousu, A. & Mattila, O. (2010). Preliminary characterization of the samples taken from a submerged arc ferrochrome furnace during operation, *Proceedings of INFACON XII 12th International Ferro Alloys Congress*, pp. 317-326, ISBN 978-952-92-7340-9, Helsinki, Finland, June 6-9, 2010.

Park, J.H.; Song, H.S. & Min, D.J. (2004) Reduction Behavior of EAF Slags Containing Cr_2O_3 Using Aluminum at 1793 K. *ISIJ International*, Vol. 44, No. 5, (May 2004), pp. 790-794.

Pelton, A. & Bale, C. (1986) A Modified Interaction Parameter Formalism for Non-Dilute Solutions. *Metallurgical and Materials Transactions A*, Vol. 17A, No. 7, (July 1986), pp. 1211-1215.

Reck, R.; Chambon, M.; Hashimoto, S. & Graedel T.E. (2010). Global Stainless Steel Cycle Exemplifies China's Rise to Metal Dominance. *Environmental Science & Technology*, Vol. 44, No. 10, (May 2010), pp. 3940-3946.

Riekkola-Vanhanen, M. (1999). *Finnish expert report on best available techniques in ferrochromium production*, Finnish Environment Institure, ISBN 952-11-0504-6, Helsinki, Finland.

Riipi, J. & Fabritius, T. (2007) Surface Tension of Liquid Fe-N-O-S Alloy. *ISIJ International*, Vol. 47, No. 11, (November 2007), pp. 1575-1584.

Riipi, J.; Fabritius, T.; Heikkinen, E-P.; Kupari, P. & Kärnä, A. (2009) Behavior of nitrogen during AOD process. *ISIJ International*, Vol. 49, No. 10, (October 2009), pp. 1468-1473.

Rousu, A.; Mattila, O. & Tanskanen, P. (2010). A laboratory investigation of the influence of electric current on the burden reactions in a submerged arc furnace, *Proceedings of INFACON XII 12th International Ferro Alloys Congress*, pp. 303-310, ISBN 978-952-92-7340-9, Helsinki, Finland, June 6-9, 2010.

Sjöberg, P. (1994). *Some aspects on the scrap based production of stainless steels*, Kungliga Tekniska Högskolan, ISBN 91-7170-861-8, Stockholm, Sweden.

Slizovskiy, D. & Tangstad, M. (2010). The effect of potassium and zinc circulation on agglomeration of a charge in SAF, *Proceedings of INFACON XII 12th International Ferro Alloys Congress*, pp. 477-485, ISBN 978-952-92-7340-9, Helsinki, Finland, June 6-9, 2010.

Tang, K. & Olsen, S.E. (2004). Computer simulation of the equilibrium relations associated with the production of manganese ferroalloys, *Proceedings of INFACON X 10th International Ferroalloys Congress*, pp. 206-212, ISBN 0-9584663-5-1, Cape Town, South Africa, February 1-4, 2004.

Tang, Y. ; Fabritius, T. & Härkki, J. (2004). Effect of fluid flows on the refractory wear in AOD, *Proceedings of 3rd Metal Separation Technology Conference*, pp. 219-226, Copper Mountain, Colorado, United States, June 20-24, 2004.

Tilliander, A. ; Jonsson, T.L.I. & Jönsson, P.G. (2004) Fundamental mathematical modelling of gas injection in AOD converters. *ISIJ International*, Vol. 44, No. 2, (February 2004), pp. 326-333.

Virtanen, E. ; Fabritius, T. & Härkki, J. (2004). Top lance practice for refining of high chromium melt in converter, *Proceedings of SCANMET II 2nd International Conference on Process Development in Iron and Steelmaking, Volume 2*, pp. 155-164, Luleå, Sweden, June 6-8, 2004.

Wei, J-H. & Zhu, D-P. (2002) Mathematical Modeling of the Argon-Oxygen Decarburization Refining Process of Stainless Steel : Part I. Mathematical Model of the Process. *Metallurgical and Materials Transactions B*, Vol. 33B, No. 1, (February 2002), pp. 111-119.

Wei, J-H.; Cao, Y.; Zhu, H-L. & Chi, H-B. (2011) Mathematical Modeling Study on Combined Side and Top Blowing AOD Refining Process of Stainless Steel. *ISIJ International*, Vol. 51, No. 3, (March 2011), pp. 365-374.

Whitney, V. (2003) Physical modelling of fluid flow in electric arc furnace caused by impinging gas jets. *Ironmaking and Steelmaking*, Vol. 30, No. 3, (June 2003), pp. 209-213.

World Steel Association. (21.1.2011). World crude steel production, In: worldsteel.org, 19.9.2011, Available from:
<http://www.worldsteel.org/pictures/newsfiles/2010%20statistics%20tables.pdf>

Wranglén, G. (1985). *An Introduction to Corrosion and Protection of Metals*, Chapman and Hall, ISBN 0-412-26050-6, New York, USA.

Xiao, Y.; Holappa, L. & Reuter, M.A. (2002) Oxidation State and Activities of Chromium Oxides in $CaO-SiO_2-CrO_x$ Slag System. *Metallurgical and Materials Transactions B*, Vol. 33B, No. 4, (August 2002), pp. 595-603.

Xiao, Y.; Schuffenger, C.; Reuter, M.; Holappa, L. & Seppälä, T. (2004). Solid state reduction of chromite with CO, *Proceedings of INFACON X 10th International Ferro Alloys Congress*, pp. 26-35, ISBN 0-9584663-5-1, Cape Town, South Africa, February 1-4, 2004.

Zhang, Y. & Fruehan, R. (1995) Effect of the Bubble Size and Chemical Reactions on Slag Foaming. *Metallurgical and Materials Transactions B*, Vol. 26B, No. 4, (August 1995), pp. 803-812.

Zhao, B. & Hayes, P.C. (2010). Effects of oxidation on the microstructure and reduction of chromite pellets, *Proceedings of INFACON XII 12th International Ferro Alloys Congress*, pp. 263-273, ISBN 978-952-92-7340-9, Helsinki, Finland, June 6-9, 2010.

Zhu H-L.; Wei J-H.; Shi G-M.; Shu J-H.; Jiang Q-Y. & Chi H-B. (2007) Preliminary Investigation of Mathematical Modeling of Stainless Steelmaking in an AOD Converter: Mathematical Model of the Process. *Steel Research International*, Vol. 78, No. 4, (April 2007), pp. 305–310.

Part 4

Heat Treatment – Modeling to Practice

5

The Estimation of the Quenching Effects After Carburising Using an Empirical Way Based on Jominy Test Results

Mihai Ovidiu Cojocaru, Niculae Popescu and Leontin Drugă
"POLITEHNICA" University, Bucharest,
Romania

1. Introduction

The graphical and analytical solutions to solve the information transfer from the Jominy test samples to real parts are shown. The essay regarding the analytical solutions for the information transfer from the Jominy test samples to real parts includes detailed information and exemplifications concerning the essence and using the Maynier-Carsi and Eckstein methods in order to determine the quenching constituents proportions corresponding to the different carbon concentrations in carburized layers, respectively the hardness profiles of the carburized and quenched layers. In the final of the chapter, taking into account the steel chemical composition, the geometrical characteristics of the carburized product, the quenching media characteristics, the heat and time parameters of the carburising and the correlations between these values and the Jominy test result, an algorithm to develop a software for the estimation of the quenching effects after carburising, based on the information provided by Jominy test, is proposed.

2. The particularities of quenching process after carburising

The aim of the quenching process after carburizing is to transform the "austenite" with high and variable carbon content of the carburized layer in quenching "martensite", respectively the core austenite in non martensitic constituents (bainite, quenching troostite, and ferrite-perlite mixture). This goal is achieved by transferring the parts from carburizing furnace into a cooling bath containing a liquid cooling (quenching) medium. The transfer can be directly made from the carburizing temperature (direct quenching), or after a previous pre-cooling of parts from the carburizing temperature to a lower quenching temperature (direct quenching with pre-cooling). In both ways, the austenitic grain size is the same (depending on the chosen carburising temperature and time), but the thermal stresses are different, being higher in the case of direct quenching and lower in the case of direct quenching with pre-cooling, due to higher thermal gradient achieved in the first cooling variant. Consequently, the risks of deformation or cracking of the parts are lower in the pre-cooling quenching, this variant being most commonly used in the industrial practice.

On the other hand, the result of quenching is influenced by three factors: **one internal, intrinsic hardenability of steel** (determined by its chemical composition - carbon content, alloying elements type and percentage) and **two external** (technological) - **thickness of the parts**, expressed by an equivalent diameter D_{ech} (the actual diameter in the case of cylindrical parts, or the diameter calculated using empirical relations for the parts with non cylindrical shapes) and **cooling capacity of quenching media**, expressed by relative cooling intensity - H (in rapport with a standard cooling media - still or low agitation industrial water at 20°C). In Fig. 1 an empirical diagram of transformation of non cylindrical sections (prisms, plates) in circular sections with the equivalent diameter D_{ech} is shown; in Table 1, the indicative values of the relative cooling intensity of water and quenching oils - H are given depending on their degree of agitation related to the parts that will be quenched. If the parts have hexagonal section, it shall be considered that the cylindrical equivalent section has the D_{ech} equal to the "key open" of hexagon.

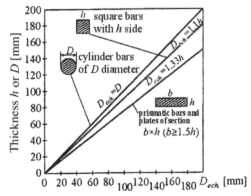

Fig. 1. Diagram for equalization of the square and prismatic sections with circular sections with diameter of D_{ech}.

Quenching media	Relative agitating degree parts/cooling medium	Relative cooling intensity, H
Mineral oils, at T=50~80°C	without agitation	0.20
	low	0.35
	average	0.45
	good	0.60
	strong	0.70
Water at approx. 20°C	without agitation	0.9
	low	1.0
	average	1.2
	good	1.4
	strong	1.6
NaCl aqueous solution, T=20°C	low	1.6
	average	2.0
	good	3.0
	strong	5.0

Table 1. Correlation between nature, degree of agitation and relative cooling intensity of common cooling media.

Lately, in the industrial practice, the so-called synthetic quenching media with cooling capacity that can be adjusted in wide limits have also been used, from the values specific to mineral oils to those specific to water, by varying the chemical composition, temperature and degree of agitation.

The degree of agitation of quenching media can be adjusted by the power and / or frequency of propellers or pumps type agitators, mounted in the quenching bath integrated in the carburizing installation (batch furnaces).

The external factors (D_{ech}, H) determine cooling law of the parts, respectively cooling curves of the points from surface or internal section of the parts; the internal factor (steel hardenability) determines quenching result, expressed by structure obtained from transformation of continuously cooled austenite from austenitizing temperature to final cooling temperature of assembly - parts-quenching medium.

To foresee or verify the structural result of quenching, the overlapping of the real cooling curves (determined by external factors D_{ech} and H) over the cooling transformation diagram of austenite of chosen steel at continuous cooling can be made, which is a graphical expression of intrinsic hardenability of steel.

The diagram of austenite transformation at continuous cooling allows steel to achieve both a quantitative assessment of the quenching structure, the estimation sizes being the proportions of martensitic and non martensitic constituents and to estimate the hardness of quenching structure.

3. Use of Jominy frontal quenching sample for estimation of quenching process results

The estimation of the steel quenching effects represents an extremely complex stage due to large number of variables that influence this operation, respectively: steel chemical composition, austenitizing temperature in view of quenching, the parts thickness and the quenching severity of the quenching media. The problem can be solved in an empirical way using the frontal quenched test sample, designed and standardized by W. E. Jominy and A. L. Boegehold and named Jominy sample (Jominy test). The simple geometry of sample and the way of performing the Jominy test covers a large range of cooling laws, their developing in terms of coordinates T- t being dependent on the distance from the front quenched part to the end of the Jominy sample. Using these curves a series of kinetic parameters of the cooling process can be obtained: cooling time and temporary (instantaneous) cooling speeds or cooling speeds appropriate for different thermal intervals. From its discovery (1938) until present, the Jominy test has been the object of numerous determinations and interpretations, evidenced especially by means of drawing of the cooling curves, of the points placed at certain distances from the frontal quenched end. The European norm, ISO/TC17/SC7N334E, Annex B1, specify the aspect of the Jominy samples cooling curves in the surface points placed at the distances dj=2.5; 5; 10; 15; 25; 50 and 80 mm from the frontal quenched end (Fig.2 [1]). This representation has the advantage that can be applied to each steel and for each austenitizing temperature T_A in terms of quenching, in the common limits T_A=830~900°C, because has on the ordinate axis the relative temperature $\theta=T/T_A$, respectively the ratio between the current temperature T (in a point placed at the

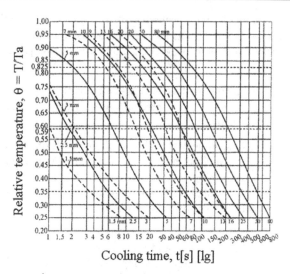

Cooling time, t[s] [lg]

Fig. 2. Cooling curves of some points placed at different distances from the frontal quenched end of Jominy sample: continuous lines _____ according to ISO/TC17/SC7N334E Standards; dashed lines - - - according to G. Murry[1].

distance d_j, after the time t) and the austenitizing temperature T_A. In the diagram presented in Fig. 2 the cooling curves for d_j=1.5; 3; 7 and 13 mm taken from the work [1] and adapted to ordinate $\theta=T/T_A=T/850$ are also shown. The most used **kinetics parameters** from the data given by the cooling curves, specific to Jominy sample and that can indicate, in a large measure, the structural result of the quenching process after carburizing are as follows:

1. *the time set for passing through the temperature of 700°C* ($t_{700} \equiv t_{0.825TA}$) and the corresponding actual cooling speed , $v_{0.825T_A} = \dfrac{T_A - 0.825T_A}{t_{0.825T_A}} = \dfrac{0.175T_A}{t_{0.825T_A}}$ for each austenitizing temperature that provides the avoiding of the transformation of the under cooled austenite in perlitical stage constituents;

2. *the time set for passing through the temperature of 500°C* ($t_{500} \equiv t_{0.59TA}$) and the corresponding actual cooling speed, $v_{500} = \dfrac{300}{t_{500}}$ for T_A=850°C, respectively $v_{0.59T_A} = \dfrac{T_A - 0.59T_A}{t_{0.59T_A}} = \dfrac{0.41T_A}{t_{0.59T_A}}$, that provides the avoiding of the transformation of the under cooled austenite in bainitic stage constituents; the thermal interval is also noted as $\Delta_{t_{0..59T_A}}^{T_A}$ or with t $_{A/5}$

3. *the time set for passing through the temperature of 300°C* ($t_{300} \equiv t_{0.35TA}$) and the corresponding actual cooling speed, $v_{300} = \dfrac{850 - 300}{t_{300}} = \dfrac{550}{t_{300}}$ for T_A=850°C and

$v_{0.35T_A} = \dfrac{T_A - 0.35T_A}{t_{0.35T_A}} = \dfrac{0.65T_A}{t_{0.35T_A}}$ for each T_A determines the passing of the under cooled austenite through the Ms ≤300°C and its transformation in quenching martensite;

4.   *the time interval* $\Delta t_{300}^{850} = t_{300} - t_{850}$ for T_A=850°C, respectively $\Delta t_{0.35T_A}^{0.825T_A}$, for each T_A and the average cooling speed of austenite in these temperature intervals ,

$$\bar{v}_{300}^{850} = \frac{850 - 300}{\Delta t_{300}^{850}} = \frac{550}{\Delta t_{300}^{850}},$$ for T_A=850°C, respectively $\bar{v}_{0.35T_A}^{0.825T_A} = \frac{0.825T_A - 0.35T_A}{\Delta t_{0.35T_A}^{0.825T_A}}$ for each T_A.

These parameters provide the cooling of the under cooled austenite in the range of M_S-M_f and its transformation in quenching martensite. The above mentioned kinetic parameters are determined for each cooling curve, corresponding to a certain distance d_J [mm] from the frontal quenched end of the Jominy test sample (Fig. 3).

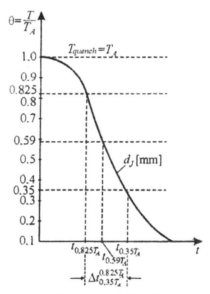

Fig. 3. The graphical determination of the temporal kinetic parameters related to the cooling curve of the points from the surface of the Jominy test sample, placed at the distance d_J [mm] from the frontal quenched end. Obs. Example for graphical determination of the temporal kinetic parameters $t_{0,825TA}$; $t_{0.59T_A} = \Delta_{0.59T_A}^{T_A} = t_{A/5}$; $t_{0.35TA}$ and $\Delta t_{0.35T_A}^{0.825T_A}$, from the cooling curve corresponding to the distance d_J[mm] from the frontal quenched end.

Analyzing the cooling curves from Fig. 2, a difference between those taken from the ISO standard and those from G. Murry work, is observed. To clarify these discrepancies and to adopt an unique and argued assessment of the way of setting out of the cooling curves in the case of Jominy samples, we can start to analyze the modality in which the heat transfers from Jominy sample to the ambient, during cooling of the sample from the austenitizing temperature T_A to the ambient temperature T_{amb} took place.

In principle, the heat flow in a point P of the Jominy sample, placed at distance x from the frontal quenched end and at the distance r from the axis of the sample, at time t after the start of cooling is given by the differential equation:

$$Q_{P(t)} = Q_{(r,x,t)} = \frac{\lambda}{r}\frac{\partial}{\partial r}\left(r\frac{\partial T_{(x,r,t)}}{\partial r}\right) \tag{1}$$

this can be solved in the following univocity conditions:
a.   initial condition: $T_{(x,0)} = T_A$;
b.   boundary conditions of first order: $T_{(0,t)} = T$ water jet (on the water cooled surface)
                                                           $T_{(x,t)} = T_{amb}$ (along the cylinder generator)
c.   boundary conditions of second order defined by the specific heat flux through the frontal cooled surface and through the external cylindrical surface cooled in air, which are proportional with the negative temperature gradients:

$$W_F = -\lambda\frac{\partial T_{(0,t)}}{\partial x} \text{ , respectively } W_{cil} = -\lambda\frac{\partial T_{(x,t)}}{\partial x} \tag{2}$$

The heat loss during sample cooling takes place by means of three mechanisms:
-   conduction – at the contact interface between cooling water and direct cooled surface, the heat loss value being a function of time:

$$W(t) = q(t) \tag{3a}$$

-   convection - of the ambient air, the heat loss value being a function of the $T_{(x,t)}$-T_{amb} difference

$$W(t) = \alpha\left[\left(T_{(x,t)} - T_{amb}\right)\right] \tag{3b}$$

where α is the convection heat transfer coefficient:
-   radiation - from the cylindrical surface:

$$W(t) = \beta\left[\left(T_{(x,t)}^4 - T_{amb}^4\right)\right] \tag{3c}$$

where β is the radiation constant, $\beta = \varepsilon\sigma$;
            ε is the emissivity coefficient of the sample surface;

$$\sigma = 5.67.10^{-8}\left[\frac{J}{m^2 sK^4}\right]$$

The solving of the differential equation (1) lead to a solution representing the general form of the cooling curves equations of points placed at the distance x from the frontal quenched end:

$$T_{(x,t)} = T_{amb} + (T_A - T_{amb})\exp\left(-\frac{c}{x}t\right) \tag{4}$$

where the parameter c has speed dimensions (m/s), and the rapport c/x is a constant on which the temporary (instantaneous) cooling speed has a linear dependency:

$$v = \frac{\partial T}{\partial t} = \left(T_{amb} - T\right)\frac{c}{x} \tag{5}$$

To simplify the analysis, without introducing further errors, can be admitted that $T_{amb} \sim 0$ and noting $T_{(x,t)} = T_S$ (the current surface temperature) and $T_S/T_A = \theta$ (the relative surface temperature), the final solution can be written as:

$$\theta = e^{-\frac{c}{x}t} \tag{6}$$

The solution (6) makes the direct connection between the relative temperature θ and time t, values that represent the coordinates in which are drawn the cooling curves of the points (planes) placed at the distance x from the frontal quench end of the Jominy sample.

In the work [2] the using of the relation (4) is exemplified in the case of Jominy samples austenitized at $T_A = 1050 °C$, for which the cooling curves of the points placed at the distances x=1, 10, 20 and 40 mm from the frontal quench end (Fig.4) were drawn and on which the ordinate referring to the relative temperature $\theta = T/1050$ and also the ISO and Murry cooling curves for distances x = 1.5 mm (Murry), 9mm (Murry),10 mm (Murry), 20 mm (ISO and Murry) and 40 mm (Murry) were also drawn.

Using data taken from continuous curves presented in Fig.4 and replacing the notation x which represents the distance from the frontal quenched end of Jominy sample with E, the value of parameter c (from eq. 6) has found as c = 0.28, so that eq.(6) of cooling curves will get a particular form (7) and the inverse function, t = f (θ) will have the expression (eq.8)

$$\theta = e^{-\frac{0.28}{E}t} \tag{7}$$

$$t = -8.244E \lg \theta \tag{8}$$

On the other hand, from Fig. 4 it can be seen that between the aspect of the actual cooling curves experimentally determined by Murry and ISO and that obtained by calculation, using the relations (7) and (8), there are differences which are substantially and simultaneously amplifying with the decreasing of the relative temperature θ. In conclusion, we can say that the theoretical analysis presented above is incomplete in that it fails to consider some effects of interaction between the types of heat loss during cooling of the Jominy samples.

Taking into account the higher matching of the Murry curves to theoretical curves, were mathematically processed the data provided by Murry curves and was noted that these are best described also by an exponential function, having the general form $t = aE^b$, and the particular form as:

$$t = aE^{1.41} \tag{9}$$

where the parameter a depends on θ also by means of an exponential relation:

$$a = 0,2\theta^{-2.4} \tag{10}$$

In conclusion, the relations describing the dependencies $t = f(\theta, E)$ - eq.(8) and $\theta = f(t, E)$ - eq. (7), will have the following particularly forms:

$$t_{(E,\theta)} = 0.2\theta^{-2.4}E^{1.41} \qquad (11)$$

$$\theta_{(E,t)} = 0.51E^{0.588}t^{-0.416} \qquad (12)$$

Once the Jominy sample kinetic parameters are known for a sample made from a certain charge of steel, they can be attributed to a specific part (with an equivalent diameter, D_{ech} known) made from the same charge of steel, taking into account that both the part and also the Jominy test sample to be processed in the same technological conditions (same austenitizing temperature T_A and time t_A) and same quenching media (with the same relative cooling intensity, H). This condition will be accomplished in the case where the Jominy sample is "embedded" in the heat treatment charge, composed of identical parts and follows the same processing sequence, in the same heating and cooling equipment. The correlation between the Jominy test sample and real part with the equivalent diameter of D_{ech} is usually graphically provided: a first chart was built by Jatczak [3] (for parts with equivalent discrete diameters of 12.5 mm, 19 mm, 25 mm, 38 mm, 50 mm and 100 mm).

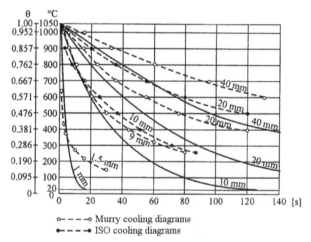

o – – – –o Murry cooling diagrams
•– – – –• ISO cooling diagrams

Fig. 4. The cooling curves drawn by means of relation (6), at distances x=1, 10, 20 and 40 mm after the cooling from T_A=1050°C of Jominy samples (continuous lines curves) and the cooling curves drawn according to ISO and Murry (dashed curves).

From this diagram in Fig. 5 (in view of illustrating of application way) the graphic for the part with D_{ech}=38 mm was displayed.

Jatczak diagram provides a graphical solution for the function dj = f (Dech, H, r), $0 \le r \le$ where R is the coordinate position of the point on the part cylindrical section, that has the same cooling law (curve) with that of points situated in the plane placed at the distance dj from the frontal quenched end of the Jominy sample.

For parts subject to carburizing, the correlation diagram - cylindrical part - Jominy sample will become the curve from the Jatczak diagram, referring to part surface S (as the layer thickness δ is very small compared to the equivalent radius of the part). In this case, the correlation function has the form dj = f (D, H) or D = f (dj, H), where D is the diameter of

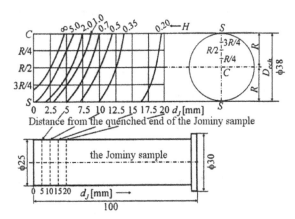

Fig. 5. Jatczak diagram of the sample (part) with D_{ech} = 38 mm and the correlation between points in section of part and points of the generator of Jominy sample placed at the distance dj from the frontal quenched end, after cooling in quenching media with different intensities of cooling., H.

cylindrical part with length L ≥ 3D (or equivalent diameter of the parts with other forms of the section) and dj is the distance from the frontal quenched end of Jominy sample.

In Fig.6,the correlation diagram- Jominy sample – superficial layer of cylindrical parts drawn by U.Wyss [4] based on the method of Grossmann and numerous literature data is reproduced.

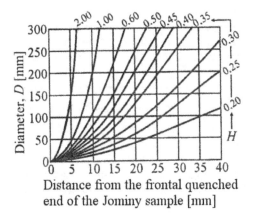

Fig. 6. The correlation curves for comparable cooling conditions in the Jominy sample and in the superficial layer of cylindrical parts at different values of cooling intensity of cooling media.

Replacing dj with notation E (the equivalent distance from the frontal quenched end of the Jominy sample) and statistically processing the data from Fig. 6, U. Wyss found the analytical relations for the inverse functions E = f (D, H) and D=f(E,H), as it follows :

- For a cooling intensity of H=0.25, corresponding to low agitated oil:

$$D_{0.25} = E^{(1.27+0.0042E)} \quad E_{0.25} = D^{(0.755-0.0003D)} \tag{13}$$

- For a cooling intensity of H=0.35, corresponding to average agitated oil:

$$D_{0.35} = E^{(1.52+0.0028E)} \quad E_{0.35} = D^{(0.649-0.0001D)} \tag{14}$$

- For a cooling intensity of H=0.45, corresponding to intensive agitated oil:

$$D_{0.45} = E^{1.755} \qquad E_{0.45} = D^{0.57} \tag{15}$$

- For a cooling intensity of H=0.60, corresponding to strong agitated oil:

$$D_{0.60} = E^2 \qquad E_{0.60} = D^{0.50} \tag{16}$$

- For a cooling intensity of H=1.00 , corresponding to low agitated water:

$$D_{1.00} = E^{(2+0.03E)} \qquad E_{1.00} = D^{(0.47-0.00015D)} \tag{17}$$

The values of cooling intensity, H, shown above, are suitable for quenching oil used at normal temperatures (50 ~ 80 °C) and with different degrees of agitation (of oil and/or parts) in a relatively wide range, starting from absence of agitation (H=0.25) to strong agitation (H=0.6), or a shower or pressure jet oil (H = 1.00), this last situation being equivalent for low agitated cooling water at 20 ° C, .

With these observations, the graphical relation between the carburized layer of part with diameter D and points from the Jominy sample, carburized in the same conditions, in accordance with the scheme shown in Fig.7 is achieved, where both the superficial layer of part and also the Jominy sample are represented at very high magnification in rapport with actual size (the dimension of the carburized layer δ~0.5~2mm and the equivalent distance in the Jominy sample E ~ 2 ~ 20mm).

4. The graphical solving of correlation between real parts - Jominy sample

From the graphical representation shown in the Fig. 7 results that the A', B' and C' points of the carburized layer of the Jominy test sample, equivalent to A,B and C points of the part carburized layer, are located on the intersections of the horizontal plane placed at the E distance from the frontal cooling plane of the Jominy test piece with the vertical planes placed at the O, δ_{ef} and δ_{tot} from the Jominy test sample generator, characterized through the C_s, C_{ef} and C_m constant carbon contents.

Based on the above considerations, a graphical solution of the issue regarding the correlation between the Jominy test sample and a part with D diameter, both carburized in identical conditions (same carbon profile and same hardness profile in the carburized layer) has developed by U. Wyss. Using of Wyss graphics solution requires the knowledge of the equivalent diameter of the part (D), the cooling intensity of quenching media (H), the carbon profile of carburized layer and the hardenability curves, respectively the hardness = f (%C) curves at various depths $0 \le \delta \le \delta_{tot}$ in the case hardened layer of the Jominy test sample.

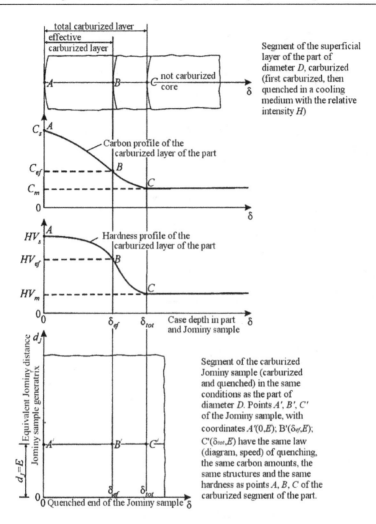

Fig. 7. The correlation scheme between part with equivalent diameter D and the Jominy sample, carburized in identical conditions.

The scheme of Wyss's graphical solution is represented in detail in Fig.8 for certain parts with the equivalent diameter D = 35 mm, made of case hardened steel with composition 0.16% C, 1%Mn and 1%Cr that were carburized at C_s = 0.8% and quenched in hot oil with average agitation (cooling intensity of H = 0.35).

The information sources used by Wyss in developing of the scheme shown in Fig. 8 were the following:

- the D=f(dj) dependence for H=0.35, has been taken from Fig. 6;
- the carbon profile curve has been experimentally plotting by means of sequential corrections of the case hardened layer of a part with dimensions of Φ35 x105 mm;

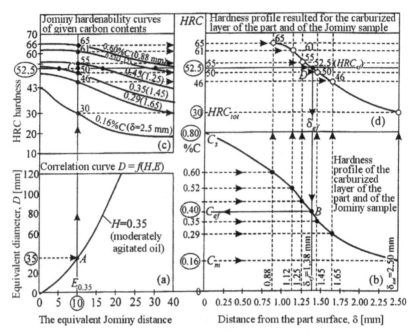

Fig. 8. Deduction of the hardness profile in the carburizing layer for parts(d) and for the Jominy test sample from Jominy hardenability curves (c) for a given carbon profile of carburizing layer (b), after carburising at C_s = 0.8% C of parts made of a case hardening steel with composition (0.16% C, 1% Mn, 1% Cr) and diameter D = 35 mm and subsequent quenching in oil with cooling intensity H = 0.35 (according to Wyss [4]).

-   the hardenability curves has been experimentally plotting by means of measuring of the HRC hardness on planes parallel with the Jominy test piece generator, corrected at the depths of the case hardened layer at which the carbon content is that mentioned on curve (0.60%C; 0.52%C; 0.45%C; 0.35%C; 0.29%C and 0.16%C).

The algorithm of using of the graphical solution is shown by arrows in Fig. 8 and involves the following steps:

-   determination of the equivalent Jominy distance E = f (D, H) - in chart (a) for D = 35 mm and H = 0.35 (the example in discussion), its value is E = 10 mm;
-   by means of vertical E = 10 mm (from the chart a) extended in the diagram (c) at its intersection with the hardenability curves, drawn for different concentrations of carbon in the case hardened layer of Jominy test sample, the appropriate hardness will be determined;
-   from the intersection points of vertical line E = 10 mm with the hardenability curves determined for different carbon concentrations in the case hardened layer of Jominy test sample are drawn horizontal lines which are extending in the space of the diagram shown in chart (d), so as to intersect the hardness profile curves for the case hardened layer (in the part and in the Jominy test sample);

- in the diagram (b) the horizontals lines corresponding to the hardenability curves from the diagram(c) will be plotted and the points of intersection with the carbon profile curve will be determined; from these points, vertical lines extending in the space of the diagram (d) are drawn and cross the horizontals plotted in an earlier stage (from the intersection points of the vertical E = 10 mm (diagram (a)) with the hardenability curves drawn for different carbon concentrations in the case hardened layer (diagram (c)); the intersection points belong to the hardness profile curve available for the existing space diagram (d). The point D in the diagram (d) in which the hardness profile curve crosses the horizontal corresponding to HRC_{ef} hardness has the abscise corresponding to the effective depth δ_{ef} (in the example discussed, for HRC_{ef} = 52.5 results δ_{ef}=1.38 mm);
- from the crossing point D, the vertical line which will meet the carbon profile curve at point B of which horizontal corresponds to the actual carbon content, C_{ef} (for example the analysis made for δ_{ef} = 1.38 mm results C_{ef} = 0.4% C).

5. Essay regarding the analytical solving of the real parts-Jominy sample correlation

The above graphical solution can be transformed into an analytical solution if the equations of the following curves are available:

a. E = f (D, H) curve;
b. C = f (δ) carbon profile curve;
c. Jominy hardenability curve, HRC = f (C, d)

a. For the equation (a) it can be started from the mathematical processing of the Wyss particular relations no (13) ~ (17) in order to their generalization. The performed attempts lead to two different types of generalized relations, applicable with satisfactory accuracy on different values ranges for the H, D and E variables:

- for $0.25 \leq H \leq 0.60$, $3 \leq E \leq 18$ (mm) and $4 \leq D \leq 50$ (mm):

$$D = \exp\left(1.5E^{0.625} - \frac{0.16 + 0.08E}{H}\right) \tag{18}$$

$$E = D^{(0.755-0.0003D)} - D\frac{H - 0.25}{3H - 0.25}; \tag{19}$$

- for $0.6 \leq H \leq 1,00$; $2 \leq E \leq 12$(mm) and $10 \leq D \leq 100$ (mm):

$$D = E^{\left[2+0.075E(H-o,6)\right]} \tag{20}$$

$$E = D^{\left[0,545-0,075H-0,000375D(H-0,6)\right]} \tag{21}$$

The parts that will be case hardened by instillation and pyrolysis of organic liquids are usually thin pieces with $D_{ech} \leq 50mm$, which are cooled in mineral oil with cooling intensity of H ~ 0.25 ~ 0.60. To this category of products it can be applied the above mentioned relations no. (18-19) thus achieving results very close to those achieved using Wyss relations (Fig.9).

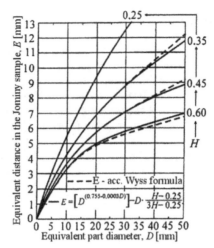

Fig. 9. Comparison between the effects of using of the Wyss particular relations and the relation no.(19) [5].

b. For the equation of the carbon profile curve (b), can be used the expression of criterial solution of the diffusion equation obtained through solving in the boundary conditions of III order – ec. 23, written under the form:

$$C_\delta = C_m + \theta(C_s - C_m) \tag{22}$$

where C_δ represents the carbon content measured at the δ depth in rapport with the surface at the case hardening end; C_m represents the carbon content of the non – case hardened core and C_s is the surface carbon content at the end of case hardening.

$$\theta = \frac{C_\delta - C_m}{C_S - C_m} = erfc\left(\frac{\delta}{2\sqrt{Dt_k}}\right) - \exp\left(h^2 Dt_k + h\delta\right).erfc\left(\frac{\delta}{2\sqrt{Dt_k}} + h\sqrt{Dt_k}\right) \tag{23}$$

where t_k represents the carburising time, $h=K/D$, represents the relative coefficient of mass transfer, K- is the global coefficient of mass transfer in the case hardening medium; D - the diffusion coefficient of carbon in austenite.

c. For the Jominy hardenability curbes (c) have been deduced by E. Just [6] several regression equations having the general form:

$$J_d = A\sqrt{C} - Bd^2\sqrt{C} + Dd - E\sqrt{d} + \sum k_i c_i - FN - G \tag{24}$$

where C is the carbon content of steel, [% C], c_i- content of alloying elements in steel [% E], d - distance from the quenched end of the Jominy test sample [mm], N - the ASTM score of the austenitic grain; A, B, D, E, F and G are the regression coefficients and J_d the HRC hardness at the d distance in the Jominy test sample.

Replacing the d distance with the equivalent Jominy distance E and J_d with $HRC_{(E)}$, can be retained the following three Just formulas, used also in other literature [4] - [7]:

$$HRC_{(E)} = 98\sqrt{C} - 0.01E^2\sqrt{C} + 1.79E - 19\sqrt{E} + (19Mn + 20Cr + 6.4Ni + 34Mo) - 7 \qquad (25)$$

$$HRC_{(E)} = 88\sqrt{C} - 0.0053E^2\sqrt{C} + 1.32E - 15.8\sqrt{E} + (5Si + 16Mn + 19Cr + 6.3Ni + 35Mo) - 0.82N - 2 \ (26)$$

These relations are valid for the following limits of carbon and alloying elements concentrations: 0.08%≤C≤0.56%; Si≤3.8%; Mn≤1.88%; Cr≤1.97%; Ni≤8.94%; Mo≤0.53% and for an austenitic score in the limits 1.5≤N≤11

$$HRC_{(E)} = 102\sqrt{C} + 1.102E - 15,47\sqrt{E} + (21Mn + 22Cr + 7Ni + 33Mo) - 16 \qquad (27)$$

relation valid for the steels with 0.25≤%C≤0.60 and with the admissible alloying elements concentrations given by relations no. (25-26).The three relations produce results significantly closer each to another and also very close to those provided by the graphical dependencies for a series of German steels presented in the work [7]. Therefore, it was adopted for explaining of the hardenability curve the relation no. (27) is written as:

$$HRC_{(E)} = 102\sqrt{C} + 1.102E - 15.47\sqrt{E} + S - 16 \qquad (28)$$

where $\qquad\qquad\qquad S = 21Mn + 22Cr + 7Ni + 33Mo \qquad\qquad\qquad (29)$

In connection with the application of relation (ec.28), have to be specified that the $HRC_{(E)}$ may not exceed a certain maximum value, which is corresponding to the hardness of the fully martensitic structure ($HRC_{100\%M}$), dependent in turn on the carbon content of martensite (austenite which is totally transformed into quenching martensite). To calculate the maximum hardness E. Just proposed the relation:

$$HRC_{100M} = 29 + 51C^{0,7} \qquad (30)$$

which is applicable to steels with carbon content in the limits 0.1%≤C≤0.6%.

Besides relation (30), in the speciality literature are presented also other relations, including the most complex relation specified in ASTM A 255 / 1989 with the polynomial expression:

$$HRC_{100M} = 35.395 + 6.99C + 312.33C^2 - 821.74C^3 + 1015.48C^4 - 538.34C^5 \qquad (31)$$

applicable to steel with C≤0.7%.

On the other hand, in the technical literature are published more data under graphical form, where are presented the hardnesses of quenching structures, experimentally determined, depending on the carbon content and the proportion of martensite in the hardening structure (Table 2). The charts from where the data from Table 2 were taken, suggest that the hardness of the quenching structures increases with carbon content in steel after curves that have the tendency to be closed to some limits values at high carbon contents. Starting from this observation, it was considered that the most appropriate analytical expression of the hardness dependence on carbon content can be obtained by using a prognosis function with tendency of "saturation".

% martensite	Quoted ref.	HRC hardness, at carbon content, %							
		0.1	0.2	0.3	0.4	0.5	0.6	0.7	0.8
100M	[a]	38.5	44	50.5	56	60	62.5	64.5	66
	[b]	37	44.5	51	56	61	63.5	65	66.5
	[c]	36	45	52	57	60	62.5	64.5	66
	[d]	36	44	50	55	59	63	65	66.5
	[e]	38.5	44.2	50.3	56.1	60.9	64.2	64.8	-
	[f]	39	45.5	51	55.8	60.4	64.6	-	-
	ref.average	37.5	44.5	50.8	56	60.2	63.4	64.7	66.2
95M	[d]	33	40	47	52.5	57	61	-	-
90M	[d]	31	38	44.5	50.5	54.5	57.5	-	-
80M	[d]	28	35	41	46.5	50.5	54.5	-	-
50M	[a]	-	32.5	36	41.5	47	51.5	53	54
	[e]	26.2	30.7	37.5	42.4	46.6	50.7	53	-
	Average of ref.	26.2	31.6	36.7	42	46.8	51.1	53	54

[a] Hodge -Orehovski(average diagram);[b] Boyd-Field; [c] G.Krauss; [d] Metals Handbook; [e] ASTM 255(relation (31)); [f]E. Just(relation(30))

Table 2. The hardness of quenching structures according to different references.

In this purpose, the data from the Table 2 were used and exponential, logharitmic and logistic functions were explored, their graphics having the ordinate at origin different from zero and positive (is known the fact that the technical iron can be quenched to a structure of „massive" acicular ferrite, close to the martensite with low carbon and which, according to relation (31) has the value 35.395 HRC. Among the explored functions, the most closed results to the results given in the technical literature, referring to the hardness of the complete martensitic structure, have been obtained with the logistic function:

$$y = \frac{K}{1 + ae^{-bx}} \tag{32}$$

and with the Johnson function:

$$\ln y = K - \frac{a}{b+x} \tag{33}$$

and their principle graphics being shown in Fig. 10.

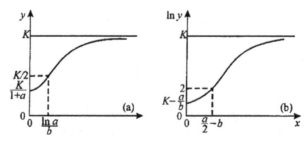

Fig. 10. The principle graphics of the logistic function (a) and Johnson function (b).

For the hardness of the complete martensitic structure, the two functions mentioned above have the particular forms:

$$HRC_{100M} = \frac{70}{1 + 1.35\exp(-4,24C)} \tag{34}$$

$$HRC_{100M} = \exp\left(4.55 - \frac{0.36}{0.27 + C}\right) \tag{35}$$

The formulas no. (34) and (35) give results very close each to another and also close to the experimental data referring to steels with carbon content in the limits 0.1~0.8%.

Furthermore the Johnson function can be used with satisfactory results also for the calculation of the hardness of the quenched layers in which the martensite proportion decreases to 50%.

The general calculation relation and the auxiliary relations are shown in Fig.11.

Whereas in many literature works the hardness is expressed in Vickers units, is necessary also a Rockwell-Vickers equivalence relation. In this purpose, mathematical tables and graphics equivalence HRC-HV - were processed and the following relation have been obtained, depending on the load used to determine the Vickers hardness:

- for loads F≤1Kgf (9,8N)
a. $HRC = 193\lg HV - 21.4\lg^2 HV - 316$

b. $\lg HV = \dfrac{193 - \sqrt{10200 - 85.6HRC}}{42.8}$ $\tag{36}$

- for loads F≥5Kgf(49N)
a. $HRC = 144.2\lg HV - 12.26\lg^2 HV - 252$

b. $\lg HV = \dfrac{144.2 - \sqrt{8436 - 49HRC}}{24.5}$ $\tag{37}$

Concerning the data provided by the graphical dependencies and relations from Fig.11 it must be specified that these are referring to the "ideal case" in which cooling of the austenite subject to quenching is achieved below M_f temperature, who, like the M_s temperature decreases with the increasing of steel carbon content (austenite) and becomes negative at higher carbon concentrations than 0.6%. In this case, if the austenite is cooled to room temperature or even above, in structure will remain a significant proportion of residual austenite, which decreases the hardness below the level indicated by the curves in Fig.11.

Typically, the proportion of residual austenite is calculated by Koistinen-Marburger relation:

$$\%A_{rez} = 100\exp\left[-0.011\left(Ms - 20\right)\right] \tag{38}$$

where the Ms point temperature can be calculated using the Brandis relation:

$$M_S = 548 - 440C - (14Si + 26Mn + 11Cr + 14Ni + 9Mo) \tag{39}$$

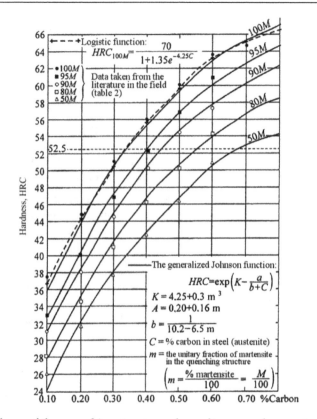

Fig. 11. The hardness of the quenching structures depending on carbon content of steel (austenite) and on the martensite proportion from the structure; continuous curves_____ drawn up with Johnsonunction; dashed curves _ _ _ _ _ drawn up with logistical function.

Decreasing of hardness caused by the presence of residual austenite can be calculated with the relation:

a. $\Delta HV \approx \dfrac{\% A_{rez}}{0.10 + 0.015(\% A_{rez})}$ (40)

derived from data provided by G. Krauss for hardnesses of fully martensitic structures (at cooling under the temperature corresponding to the end of martensitic transformation - M_f) and the structures formed by martensite and residual austenite (at cooling to room temperature). If the hardness of carburized layer is measured in Rockwell units the indicative relation can be used:

b. $\Delta HRC \approx \dfrac{\% A_{rez}}{1 + 0.2(\% A_{rez})}$ (41)

Returning to the analytical solution of the correlation between Jominy sample and real parts, which should finally allow to draw the hardness curve of the carburized and quenched layer, is noted that this solution is materialized in a mathematical model of post carburising quenching, whose solving algorithm is based on knowing of the initial data referring to the following independent variables:

a. chemical composition of steel, respectively the alloying factor;

$$S = 21Mn + 22Cr + 7Ni + 33Mo$$

b. the geometry of parts subject to carburising, respectively the equivalent diameter D_{ech};
c. the cooling intensity of the quenching medium, respectively the Grossmann (H) relative cooling intensity;
d. the requested effective hardness (HRC_{ef});
e. the requested effective case depth(δ_{ef}).

Starting from this initial data, the algorithm for determining of the hardness curve of carburized and quenched layer will require the following sequence of steps:

Step I taking into account the geometry of parts subject to processing, the equivalent diameter D_{ech} is calculated with one of the equivalence relations mentioned in Fig.1.

Step II taking into account D_{ech} and h, is calculated the equivalent distance E from the frontal quenched end of the Jominy sample by means of the relation (19) , written under the form:

$$E = D_{ech}^{(0.755-0.0003D_{ech})} - D_{ech}\left(\frac{H-0.25}{3H-0.25}\right) \qquad (42)$$

Step III taking into account E and S and giving to the hardness the value HRC_{ef} , can be calculated the effective carbon content C_{ef} by means of relation (28), written under the form:

$$HRC_{ef} = 102\sqrt{C_{ef}} + 1.1E - 15.47\sqrt{E} + S - 16 \qquad (43)$$

which lead to the relation:

$$C_{ef} = \left[\frac{(HRC_{ef}+16-S)+15.47\sqrt{E}-1.1E}{102}\right]^2 \qquad (44)$$

In the technical literature and in industrial practice of carburising followed by quenching, for the actual hardness value is used most frequently $HRC_{ef} = 52.5$ (ie $HV_{ef}=550$).Using this value, the equation (44) takes the following particular form:

$$C_{ef52.5HRC(550HV)} = \left[\frac{(68.5-S)+\left(15.47\sqrt{E}-1.1E\right)}{102}\right]^2 \qquad (45)$$

U.Wyss had drawn the curves (parabola) $C_{ef52.5HRC} = f(E,S)$ for several German case hardening steels (Fig.12) with the average chemical composition (according to DIN-tab.3), without mentioning of the calculation formula or the effective values of the alloying factors S.

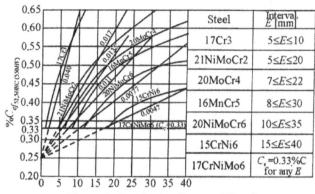

Steel	Interval E [mm]
17Cr3	$5 \leq E \leq 10$
21NiMoCr2	$5 \leq E \leq 20$
20MoCr4	$7 \leq E \leq 22$
16MnCr5	$8 \leq E \leq 30$
20NiMoCr6	$10 \leq E \leq 35$
15CrNi6	$15 \leq E \leq 40$
17CrNiMo6	$C_t = 0.33\%C$ for any E

The equivalent Jominy distance, E [mm]

Fig. 12. Correlation between the effective carbon content and the Jominy equivalent distance for different German steels with the average standard chemical composition[4].

In addition, U. Wyss suggested the using of a linear relation to calculate the actual carbon content as it follows:

$$C_{ef,52.5HRC} = 0.25 + f_c E \tag{46}$$

In which the slope f_c is a hardenability factor of steel, having the values written on the corresponding lines, drawn in Fig.12.

In connection with this approach of the problem, it is observed that the use of relation (46) is possible only for certain intervals corresponding to equivalent distance E, intervals on which the value obtained for C $_{ef52.5}$ is not lower than 0.33% and is not different more than ±0.03% in rapport with curve (parable) replaced by the line corresponding to a given steel. These limits for E are mentioned also in the table placed on the right of Fig. 12.

In the work has been calculated the effective carbon content with formula (45), both for the German steels and also for the Romanian steels having an average chemical composition according to DIN and STAS (Table 3), using for the calculation of the alloying factor the relation S=21Mn+22Cr+7Ni+33Mo. The variation of the effective carbon content with the Jominy equivalent distance for the specified steels is shown in Fig. 13.

The analysis of the curves from Fig.13 led to following conclusions:

a.   for the steels with close values of the alloying factor, S, the curves are close positioned or even overlapped (case of 20MoNi35, 16MnCr5, 18MnCr10 and 21TiMnCr12 steels, having S=44.3 ~45.25 and of the steels 15CrNi6 și 21MoMnCr12, having S~55.55);

b.   the ordinates at origin and the rate of curves are strongly decreasing with the increasing of the value of the alloying factor, S. As a result, the curves can be replaced with straight lines with different slope, but also with ordinates having different origins, both being dependent of alloying factor S. Putting the condition that the replacing lines do not lead to deviations higher than ±0,03%C, the generalized equation of these lines was:

$$C_{ef52.5} = C_{ef0} + f_c E \tag{47}$$

where the ordinate at origin C_{efo} and the slope f_c are dependent on the alloying factor S, respectively:

$$C_{ef0} = 0.41 - 0,005S \tag{48}$$

$$f_c = 6,5e^{-1.6\ln S} \tag{49}$$

Steel grade	The average chemical composition,%						Alloying factor, S
	C	Mn	Cr	Ni	Mo	Ti	
15Cr3	0.15	0.50	0.65	-	-	-	25.30
15Cr08	0.15	0.55	0.85	-	-	-	30.25
21NiMoCr2	0.21	0.80	0.50	0.55	0.20	-	38.25
20MoCr4	0.20	0.75	0.40	-	0.45	-	39.40
16MnCr5	0.16	1.15	0.95	-	-	-	45.05
18MnCr10	0.18	1.05	1.05	-	-	-	45.15
21TiMnCr12	0.21	0.95	1.15	-	-	0.07	45.25
20MoNi35	0.20	0.55	-	3.50	0.25	-	44.30
13CrNi30	0.13	0.45	0.75	2.95	-	-	46.60
20NiMoCr6	0.20	0.60	0.70	1.50	0.30	-	48.40
15CrNi6	0.15	0.50	1.55	1.55	-	-	55.45
21MoMnCr12	0.21	1.00	1.20	-	0.25	-	55.65
17CrNiMo6	0.17	0.50	1.65	1.55	0.30	-	67.55

Table 3. The average chemical compositions and the alloying factor S of the standardized German and Romanian case hardening steels.

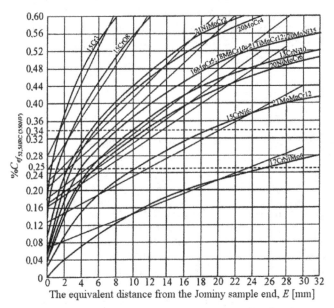

Fig. 13. The variation of the effective carbon content C $_{ef52,5}$ with E for German and Romanian steels, having the average chemical composition (in the limits of the grade).

The particular values of these elements and the applicable intervals of the linearizing relation for the standardized case hardening steels are given inTable 4.

Steel	Alloying factor,S	C_{ef0}	f_c	E interval* (%C_{ef})
15Cr3	25.30	0.28	0.043	4(0.45)~7(0.58)
15Cr08	30.25	0.26	0.030	4(0.38)~10(0.56)
21NiMoCr2	38.25	0.22	0.0195	4(0.30)~18(0.57)
20MnCr4	39.40	0.21	0.018	4(0.28)~20(0.57)
20MoNi35 16MnCr5 18MnCr10 21TiMnCr12	45	0.185	0.0145	4(0.24)~26(0.56)
13CrNi30	46.6	0.18	0.014	4(0.24)~25(0.52)
20NiMoCr6	48.4	0.17	0.013	4(0.22)~25(0.50)
15CrNi6 21MoMnCr12	55.5	0.13	0.0105	4(0.17)~30(0.44)
17CrNiMo6	67.5	0.07	0.0077	4(0.10)~30(0.30)

*For E have been taken values as E≥4mm, imposed by the E.Just rel and rel.(19)

Table 4. The values of parameters C_{ef0}, f_c and the intervals for E in which is applied the linearizing relation (47) for the German and Romanian steels with medium standardized chemical composition.

In fact, the effective carbon content can has lower values than those ensuring a certain minimum proportion of quenching martensite. If in Fig.11 will be drawn the horizontal corresponding to the effective hardness of 52.5 HRC can be found that this value is assured by the following combinations of effective carbon contents and respectively martensite percentages in the hardening structure:

%Martensite (M)	100	95	90	80	50
%$C_{ef\,52.5}$	0.34	0.40	0.45	0.56	0.70

On the interval 80≤M≤100%, the effective carbon content has a lineraly variation with the martensite proportion, according to relation:

$$\%C_{ef\,52.5} = 1,44 - 0,011M \qquad (50)$$

Setting the condition that for the effective case depth the actual proportion of martensite to be within the required range (to provide appropriate mechanical characteristics of the carburized and quenched layer), can be noted that the maximum amount of the effective carbon content should not exceed 0, 56% (value which is close to the carbon surface content of about 0.8%, with a drastic reduction of the carburizing depth, particularly in steels with low hardenability, respectively with an alloying factor S ≤ 30).On the other hand, the value of the C $_{ef52,5}$ will not decrease more below 0.34%C because even the quenching structure for the effective depth is fully martensitic, its hardness decreases significantly below the set value (52.5HRC). This is the reason for why U.Wyss adopted for the 17CrNiMo6 steel the amount C_{ef}=0.33%, although in the carburized layer of steel, the information offered by relation (45) and Fig.13 shown that the hardness of 52.5HRC can be obtained even for the content of 0.17%C of core (at E=13 mm). In a subsequent paper [7], Weissohn and Roempler suggest as minimum value for the carbon content with the concentration of 0.28% (for which HRC$_{100M}$~49.4, according to Fig.11). Adopting a value below that of the 0.34% could be justified for alloyed steel intended for carburising, due to the fact that alloyed martensite has a hardness higher with 1~2HRC than that of unalloyed steels.

In conclusion, for the calculation of the effective carbon content ($C_{ef,52.5HRC}$) can be used the linearizing relations (47-49),with supplementary restrictions:

a.   E≥4mm
b. $0,28 \leq C_{ef,52.5HRC} \leq 0.56\%$ (respectively 100≥M≥80%)

Step IV Using C_{ef} and δ_{ef} , can be calculated the carburising time (t $_k$) at isothermal carburising with a single cycle or the active carburising time (t $_k$) ,respectively the diffusion time (t$_D$), at carburising in two steps. The performing of this calculation supposes the knowing of thermal regime (t$_k$,t$_D$), the chemical regime(C_{potK},$C_{pot\ D}$), the corresponding evaluation of the global mass transfer in the carburising medium (K); and the diffusion coefficient of carbon in austenite (D).

Step V is based on knowing of the carbon profile and of the cooling law (curve) of the layer and has as final purpose the drawing of the layer hardness profile. The carbon profile can be determined after the step IV, and the cooling curve of layer can be drawn using the relation (11).

The most direct method for determining and drawing of the hardness profile consist in overlapping of the cooling curve of the case hardened layer on the transformation diagrams of continuous cooling of austenite (CCT diagrams) „of different steels" from the layer, steels with carbon content that decreases from surface to core. The method is illustrated in Fig.14, for the case where the diagrams for the austenite transformation, corresponding for three steels with different carbon content that will be carburized, are known:

a.   with carbon content of core, C_m ;
b.   with effective carbon content, C_{ef} ;
c.   with carbon content of surface, Cs

Because the temperatures corresponding to Ms point and the transformation ranges for the three diagrams are placed differently in the plane T-t (at a lower position and to the right, as the carbon content increases), the intersections of these with the cooling curve led to different structures (decreasing of the proportions of bainite and increasing of the proportions of martensite), respectively with different hardnesses (HRC$_m$< HRC$_{ef}$< HRC$_{100M}$

For an accurate drawing of the hardness profile, is necessary to know a minimum number of 5-6 austenite transformation diagrams, corresponding to different carbon contents for a steel having in its chemical composition, all the other elements that are permanently accompanying and alloying elements with the same contents. But, this kind of technical "archive" is not currently available, even in the richest databases for the usual case-hardening steels. To overcome these difficulties can be used a hardness calculation method based on knowledge of chemical composition and of a kinetic parameter characteristic to cooling law of the case hardened layer. This parameter can be the cooling time at passing through a certain temperature enclosed between the austenitizing temperature (T_A) and ambient temperature (T_{amb}). Most of the kinetic parameters of this type are the times of passing through temperatures of 700°C, 500°C and 300°C respectively t_{700}, t_{500} şi t_{300}, highlighted also on CCT diagrams in Fig. 14. Because the data in the literature in the field are referring usually to a temperature of T_A=850°C, the absolute parameters mentioned can be replaced with relative parameters $t_{0.825T_A}, t_{0.59T_A}$, respectively $t_{0.35T_A}$. Besides these cooling times can be used also other kinetic parameters as are the cooling times between two given temperatures ($\Delta t_{500}^{T_A}, \Delta_{500}^{800}, \Delta_{300}^{700}$), or the instantaneous cooling speeds at passing of the cooling curve through a temperature ($v_{750}, v_{700}, v_{500}, v_{300}$), or the medium cooling speed in a temperature range (v_{300}^{-700}). The advantage of using of these kinetic parameters is that can be built structural diagrams in coordinates T-lgt or T-lgv, in which the cooling curves are replaced by verticals lgt or lgv and on basis of these, also structural diagrams in coordinates % structural constituents – lgt (lgv).

5.1 Method Maynier-Carsi

In the works [8] and [9] is used a calculation method derived from the analysis of 251 diagrams of transformation of austenite at continuous cooling, method in where the kinetic factor taken in consideration is the instantaneous cooling speed at passing of the cooling curve through the temperature of 700°C, respectively:

$$v_{700} = \frac{T_A - 700}{t_{700}} \tag{51}$$

In the calculation, the authors have introduced also the austenitizing parameter:

$$P_A = \left[\frac{1}{T_A} - \frac{4.6}{\Delta H} \ln t_A \right]^{-1} \tag{52}$$

where T_A represents the austenitizing temperature [K], t_A is the austenitizing time [h], with condition that t_A<1h, and ΔH the activation energy of the austenitization process (ΔH=2.4.10⁵ [cal/molK] for steels with low than 0.04%Mo şi ΔH=4.2.10⁵ [cal/molK] for steels with more than 0.04%Mo). The P_A parameter determines, indirectly, the size of the austenitic grain, after the heating at T_A, in time t_A.

By means of statistical processing have been derived [5] the regression relations for ten critical cooling speeds named v_{cr}, v_{100M}, v_{90M}, v_{50M}, v_{10M}, v_{1M}, v_{1FP}, v_{10FP}, v_{50FP}, v_{90FP}, v_{100FP} in which the figures shown the martensite proportions (M), respectively ferrite and perlite (FP).

The regression relations have the following general form:

$$\lg v_{cr} = C_v - \left(\sum K_{E_i}.E_i + K_A.P_A \right) \tag{53}$$

in which C_v is the speed constant, K_{Ei} – the influencing coefficient of Ei element, P_A represents the austenitizing parameter, K_A- the influencing coefficient of austenitizing

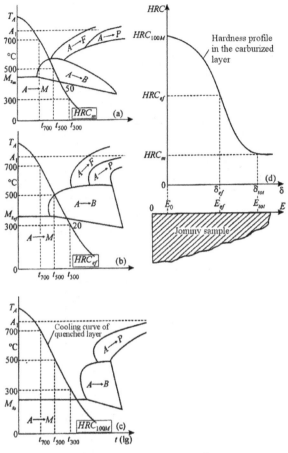

Fig. 14. Determination of the structure and hardness in different points in the carburized layer depth of the part and the Jominy sample. a) the CCT diagram of the not carburized core, with $\delta \geq d_{tot}$ and the core carbon content C_m; b) the CCT diagram of the layer point at depth def, with carbon content of C_{ef}; c) the CCT diagram of the surface point of the part ($\delta = 0$) with surface carbon content C_{ef}; d) hardness profile of the carburized layer in the part and the Jominy sample.

parameter (of the austenitic grain size), and Ei represent the proportion of carbon, adding elements and alloying elements.

The particular forms of the regression relations are given in Table 5, for the case in which these cooling speeds are expressed in [°C/h]. To use this calculation method is necessary to know the instantaneous cooling speed v_{700} in the case hardened layer and its dependency of this of equivalent distance E from the frontal quenched end of Jominy sample.

In this purpose, Popescu and Cojocaru [5] have used a diagram v_{700} =f(E), taken from UNE7279 norm, having the ordinate v_{700} [°C/s] in logarithmic scale and the abscise E[mm] in normal scale (Fig.15).

$\lg v_{cr_{700}}$	C_v	K_{E_i} for the element:					K_A
		C	Mn	Ni	Cr	Mo	
$\lg v_{100M}$	9.81	4.62	0.78	0.41	0.80	0.66	0.0018
$\lg v_{90M}$	8.76	4.04	0.86	0.36	0.58	0.97	0.0010
$\lg v_{50M}$	8.20	3.00	0.79	0.57	0.67	0.94	0.0012
$\lg v_{10M}$	9.80	3.90	$-0.54\,Mn\,++2.45\,\sqrt{Mn}$	0.46	0.50	1.16	0.0020
	8.56	1.50	1.84	0.78	1.24	1.46	0.0020
$\lg v_{1FP}$	10.55	4.80	0.80	0.72	1.07	1.58	0.0026
$\lg v_{10FP}$	9.06	4.11	0.90	0.60	1.00	2.00	0.0013
$\lg v_{50FP}$	8.04	3.40	1.15	0.96	1.00	2.00	0.0007
$\lg v_{90FP}$	8.40	2.80	1.51	1.03	1.10	2.00	0.0014
$\lg v_{100FP}$	8.56	1.50	1.84	0.78	1.24	$2\sqrt{Mo}$	0.0020

Table 5. The values of C_v, K_{E_i} and K_A (P_A in °C) from the regression relations (53)

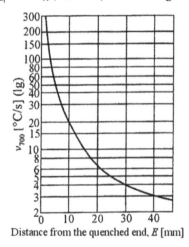

Distance from the quenched end, E [mm]

Fig. 15. Dependency of the instantaneous cooling speed V_{700} on distance from the frontal quenched end of the Jominy sample (according to UNE 7279 [9].

The mathematical processing of data from this diagram shows that the function $v_{700}=f(E)$ has the form:

$$v_{700} = 500E^{-1.41} \ [°C/s] \tag{54}$$

$$v_{700} = 1{,}8.10^6 E^{-1.41} \ [°C/h] \tag{55}$$

The two relations offer results in accordance with data of Fig. 15, for 1.5≤E≤30mm.

Admitting that the austenitizing temperature, for the data shown in Fig.15 is T_A=850°C (which is not specified in the work [9], but is usually used in other works in the field), the absolute instantaneous cooling speed v_{700} can be replaced with the relative speed v $_{0,825TA}$, which will be calculated with the relation:

$$v_{0,825T_A} = \frac{T_A - 0.825T_A}{t_{0.825T_A}} = \frac{0.175T_A}{t_{0.825T_A}} \tag{56}$$

Combining this relation with the relation (11), in which $\theta = 0.825$, results:

$$v_{0.825T_A} = 0.55T_A.E^{-1.41} \tag{57}$$

For $T_A = 850°C$, relation (57) led to $v_{700} = 468.E^{-1.41}$, value which is very close to that given by the relation(54), this is confirming also the validity of the relations (11) and (12).

Taking into account the critical speeds (calculated with the relations from Table 5) and the carbon profile curve of the carburized layer allow the overlapping of the structural diagrams at different carbon concentrations and taking into account the speed v700 (v0,825TA) of layer, allow the positioning of the vertical of this speed in the space of the structural diagram and the deriving of the proportions of the quenching constituents for each carbon content (respectively for each depth) of carburized layer.

5.2 The Eckstein method

In paper [10], another drawing method of the hardness profile curve is provided, which is considered to be described as a complex exponential function as it follows:

$$H_E = H_S.E^b.e^{-cE} \tag{58}$$

where H_S is the surface hardness, respectively the martensite hardness which has the superficial carbon content C_s, E is the equivalent Jominy distance in the depth of the case hardened layer, and b and c are the coefficients dependent on the steel chemical composition.

For the calculation of the Jominy equivalent distance, the author provides the formula:

$$E = 0.0877 + 0.761t_{A/5} - 0.0148t_{A/5}^2 + 0.00012t_{A/5}^3 \tag{59}$$

where $t_{A/5}$ represents the cooling time from austenitizing temperature in view of cooling, from 850°C, until 500°C.

The relation no.(59) is applicable for the case $t_{A/5} \leq 42s$ (maximum values for parts with the equivalent diameter D≤100mm, quenched in mineral oils with cooling intensity 0.25≤H≤0.60).

Due to fact that $T_A = 850°C$ and $T_{cooling} = 500°C$, results that the relative cooling temperature has the value $\theta_{racire} = \frac{500}{850} = 0.59$ and the time $t_{A/5}$ can be replaced with time $t_{0.59T_A}$, usable for each austenitizing temperature in view of cooling, T_A. With this specification, the relation (11) can be used, from where, through the replacement $\theta_{(E,t)} = 0.59$ and through variable changing, can be obtained the relation:

$$E = 1.24t_{0.59T_A}^{0,71} \tag{60}$$

In Fig. 16 is graphically shown the dependency $E = f(t_{0.59T_A})$ determined with relations (59) and (60), from where results that these have led to identical results for E≤25mm and very close results for the range 25≤E≤40mm.

Due to fact that in both relations, the independent variable is the cooling time $t_{0.59T_A}$, become necessary to find a modality for calculation of this time depending on part diameter , D, on cooling intensity of the quenching medium, H and on the carburising depth δ. In this purpose, were used the data provided by the graphical dependencies from work [10], referring to the experimental determination of the time $t_{A/5}$ in accordance with the carburising depth δ, which are combined with the data provided by Fig. 16 and 6 of the work. From this processing was derived, in a first and acceptable approximation, the cooling time from austenitizing temperature T_A until temperature T=0.59T_A having a linearly dependence on δ depth, respectively:

$$t_{0.59T_A} = t_0 + m\delta \ [s] \tag{61}$$

with
$$t_0 = 13.6 + 0.36D - 23.4H - 0.126DH \tag{62}$$

$$m = 0.177D^{0.575} \tag{63}$$

where D is the part diameter [mm], and H is the relative cooling intensity of the quenching medium.

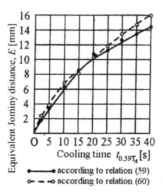

Fig. 16. The dependence of the Jominy equivalent distance and the cooling time from the austenitizing temperature T_A to the temperature of T=0,59 T_A ($t_{0.59T_A}$ or $t_{A/5}$ for T_A=850°C).

In order to properly use the relation (58), the b and c coefficients also have to be known; for these coefficients no information is available in the technical literature, including the work [10].

In conclusion, the effects of post carburising quenching process can be quantified by the calculation algorithm required by Maynier-Carsi, but corrections have to be applied; these corrections are determined by the presence of residual austenite and its presence implication on the hardness in the superficial layer; the obtained algorithm allows the very easily determination of the information regarding the effects of the carburising and quenching process on layer characteristics, starting from information provided by Jominy test.

Algorithm for developing of software dedicated to the estimation of the carburising effects

HV=f(%C);

HV=F(distance from the surface of the carburized layer)

Input data

- Steel chemical composition;
- Equivalent diameter, Dech, cm;
  Equalization relations
  Dech.=1,1h (for a square section, h is the section side)
  D circular sections
  D=1.33h for rectangular sections with b≥1,5 h;
- Cooling intensity, H;
- Effective hardness, HRC_{ef} (52.5 HRC);
- Requested effective carburized depth, δ_{ef}, cm;
- Diffusion coefficient, D, cm²/s;
- Austenitizing temperature, T_A [°C];
- Austenitizing time, t_A [h] (maximum 0.5 h);
- Carburising temperature, T_K, K;
- Carbon concentration at the surface of the carburized layer, Cs=0.8%.

STEPS

START

1° Calculates the equivalent distance from the frontal quenched end of the Jominy sample, E

$$E = Dech^{(0.755-0.0003Dsch)} - Dech\frac{H - 0.25}{3H - 0.25}[cm];$$

2° Calculates the alloying factor, S

$$S = 21Mn + 22Cr + 7Ni + 33Mo ;$$

3° Calculates the carbon effective concentration, Cef, corresponding to effective hardness

$$C_{ef\,52.5HRC(550HV)} = \left[\frac{(68.5 - S) + (15,47\sqrt{E} - 1.1E)}{10^2}\right]^2 ;$$

4° Calculates the austenitizing parameter,P_A;

| If %Mo<0,04 | If %Mo>0,04 |

$$P_A = \left[\frac{1}{T_A + 273} - 1.961.10^{-5}\ln t_A\right]^{-1} \qquad P_A = \left[\frac{1}{T_A + 273} - 1.095.10^{-5}\ln t_A\right]^{-1}$$

5° Calculates the logarithm of critical cooling speeds $\log v_{700}$ for the constituents percentages: 100%M, 90%M, 50%M, 10%M, 1%M, 1%FP, 10%FP, 50%FP, 90%FP, 100%FP and carbon concentrations between surface and core

$$\left(C \in \left[C_m; \frac{C_m + C_{ef}}{2}; C_{ef}; \frac{C_{ef} + C_S}{2}; C_S\right]\right)$$

$$V_{Cr700_{100\%M}} = 9.81 - (4.62\%C + 0.78\%Mn + 0.41\%Ni + 0.80\%Cr + 0.66\%Mo + 0.0018P_A);$$

$$V_{Cr700_{90\%M}} = 8.76 - (4.04\%C + 0.86\%Mn + 0.36\%Ni + 0.58\%Cr + 0.97\%Mo + 0.0010P_A);$$

$$V_{Cr700_{50\%M}} = 8.20 - (3.0\%C + 0.79\%Mn + 0.57\%Ni + 0.67\%Cr + 0.94\%Mo + 0.0012P_A);$$

$$V_{Cr700_{10\%M}} = 9.80 - (3.9\%C - 0.54\%Mn + 2.45\sqrt{\%Mn} + 0.46\%Ni + 0.50\%Cr + 1.16\%Mo + 0.0020P_A);$$

$$V_{Cr700_{1\%M}} = 8.56 - (1.5\%C + 1.84\%C + 0.78\%Ni + 1.24\%Cr + 1.46\%Mo + 0.0020P_A);$$

$$V_{Cr700_{1\%FP}} = 10.55 - (4.80\%C + 0.80\%Mn + 0.72\%Ni + 1.07\%Cr + 1.58\%Mo + 0.0026P_A);$$

$$V_{Cr700_{10\%FP}} = 9.06 - (4.11\%C + 0.90\%Mn + 0.60\%Ni + 1\%Cr + 2\%Mo + 0.0013P_A);$$

$$V_{cr700_{50\%FP}} = 8.04 - (3.40\%C + 1.15\%Mn + 0.96\%Ni + 1\%Cr + 2\%Mo + 0.0007P_A)$$

$$V_{Cr700_{90\%M}} = 8.40 - (2.80\%C + 1.51\%C + 1.03\%Ni + 1.1\%Cr + 2\%Mo + 0.0014P_A);$$

$$V_{cr700_{100\%FP}} = 8.56 - (1.5\%C + 1.84\%Mn + 0.78\%Ni + 1.24\%Cr + 2\sqrt{Mo} + 0.0020P_A)$$

6° Calculates critical cooling speed v_{700} and its logarithm

$$V_{700} = 1.8.10^6 . E^{-1,41} [°C / h]$$

$$\lg V_{700}$$

7° Identifies between 50 values calculated at point 5° the intervals where is the $\lg v_{700}$ critical cooling speed; interpolates the value and determines the constituents percentages;

8° Display in a centralised manner the %FP, %M and by difference, %B;

9° Calculates for different carbon percentages

$$\left[Cm; \frac{Cm + Cef}{2}, Cef, \frac{Cef + Cs}{2}, Cs\right];$$

The hardnesses of martensite, bainite, ferrite+perlite constituents;

$$HV_M = 902.6\%C + 26.68 \lg V_{700} + 121.156$$

$HV_B = 185\%C + 330\%Si + 153\%Mn + 144\%Cr + 65\%Ni + 191\%Mo + (89 + 53\%C - 55\%Si$
$-22\%Mn - 20\%Cr - 10\%Ni - 33\%Mo)\lg V_{700} - 323$

$HV_{FP} = (1329\%C^2 - 744\%C + 15\%Cr + 4\%Ni + 135.4)\lg V_{700} + 3300\%C - 5343C^2 - 437;$

10° Calculates the global hardness, $HV_{mixture}$ for different carbon concentrations

$HV_{mixture} = \%FP.HV_{FP} + \%B.HV_B + \%M.HV_M;$

11° Draw up $HV_{mix} = f(\%C);$

12° For a set value for the effective case depth, δ_{ef}, determines the carburising depth, t_K, from the relation:

$C_{ef} = C_0\left[\dfrac{0.79\sqrt{D.t_K} - 0.24\delta_{ef}}{\delta_{ef}}\right](C_S - C_0);$

13° For different values of the maintaining time at carburising, in the $(o;t_K]$, determines the correlation of $C_\delta = f(\delta)$, for $\delta \in (o, \delta_{ef}]$,

$C_\delta = C_0\left[\dfrac{0.79\sqrt{D.t_K} - 0.24\delta}{\delta}\right](C_S - C_0);$

14° Calculates Ms, for different carbon percentages, between C_0 and C_s for

$\left[C_0;\dfrac{C_0 + Cef}{2}, Cef, \dfrac{Cef + Cs}{2}, Cs\right];$

$M_s = 548 - -440\%C - (14\%Si + 26\%Mn + 11\%Cr + 14\%Ni + 9\%Mo);$

15° Calculates the proportion of residual austenite for:

$\left[C_0;\dfrac{C_0 + Cef}{2}, Cef, \dfrac{Cef + Cs}{2}, Cs\right];$

$\%AR = 100\exp[-0.011(Ms - 20)];$

16° Calculates the hardness variation due to residual austenite for different carbon proportions

$\left[C_0;\dfrac{C_0 + Cef}{2}, Cef, \dfrac{Cef + Cs}{2}, Cs\right];$

$\left[\Delta HV = \dfrac{\%AR}{0.10 + 0.015\%AR}\right];$

17° Calculates the corrected value of hardness, $HV_{corrected}$

$HV_{corrected} = HV_{am} - \Delta HV;$

18° Draw up the dependency $HV_{corrected} = f(\%C)$ for the carbon content in limits $[C_o, C_s];$

19° Corroborate the data from stages 13 and 18 and draw up $HV_{corrected} = f(\delta)$

STOP

Fig. 17. Algorithm for a software used for in characterization of the effects of the carburising-quenching cycle applied to case hardening steels.

6. References

[1] Murry, G. (1971). *Mem.Scient.Rev.Met*, no.12,pp.816-827

[2] Bussmann, A. (1999). Definition des mathem.Modells,*CET*

[3] Jatczak, C. F. (1971). Determining Hardenability from Composition. *Metal Progress*, vol.100, no.3, pp.60-65

[4] Wyss, U. (1988) Kohlenstoff und Härteverlauf in der Einsatzhärtungsschicht verschieden legierter Einsatzstähle, *Härt.Tech.Mitt.*, no.43,1

[5] Popescu, N., Cojocaru, M. (2005). Cementarea oțelurilor prin instilarea lichidelor organice. *Ed.Fair Partners*, pp.115, Bucuresti

[6] Just, E. (1986). Formules de trempabilité. *Härt.Tech.Mitt.*, no.23, pp.85-100

[7] Roempler, D., Weissohn, K. H. (1989). Kohlenstoff und Härteverlauf in der Einsatzhärtungsschicht-Zusatzmodul für Diffusionrechner, *AWT-Tagung*, Einsatzhärten, Darmstadt

[8] Maynier, Ph., Dollet, J., Bastien, P. (1978). Hardenability Concepts with Applications to Steel, *AIME*, pp.163-167, 518-545

[9] Carsi, M., de Andrés, M.P. (1990). Prediction of Melt-Hardenability from Composition. *Symposium IFHT*,Varşovia

[10] Eckstein, H.J. (1987). Technologie der Wärmebehandlung von Stahl. VEB Deutscher Verlag für Grundstoffindustrie,Leipzig

Part 5

Semi-Solid Processing and
Die Casting of Alloys and Composites

6

Semisolid Processing of Al/β-SiC Composites by Mechanical Stirring Casting and High Pressure Die Casting

H. Vladimir Martínez[1] and Marco F. Valencia[2]
[1]Institute of Energy, Materials and Environment, School of Mechanical Engineering,
Pontificia Bolivariana University, Medellín,
[2]Engineering School of Antioquia, Medellín,
[1,2]Colombia

1. Introduction

Metals and alloys are generally produced and shaped in bulk form but can also be intimately combined with another material that serves to improve their performance. The resulting material is known as a metal matrix composite (MMC). This class of composite encompasses many different materials that can be distinguished according to their base metal (e.g., aluminium, copper, titanium), their reinforcement phase (e.g., fibers, particles, whiskers), or their manufacturing process (e.g., powder metallurgy, diffusion bonding, infiltration, mechanical or electromagnetic stir casting and die casting).

Processing advantages make die casting one of the most efficient technologies available for producing a wide range of durable and rigid MMC products for use in commercial, industrial and consumer applications. There are several well-established die casting methods that can be used to produce castings for specific applications. Including squeeze casting and semisolid molding (thixocasting and rheocasting). Squeeze casting is a method by which molten alloy is cast without turbulence or gas entrapment at high pressure to yield high-quality, dense and heat-treatable components). Thixocasting is a procedure whereby semisolid metal billets, with no dendritic microstructure, are cast to provide dense, heat-treatable castings with low porosity. Rheocasting refers to several processes that allow the creation of globular structures and thixocasting involves the reheating of ingots obtained by rheocasting until the semisolid gap followed by one semisolid molding step.

Modern technology is currently geared towards net-shape processes, which are able to eliminate the intermediate storage of ingots (required in thixocasting), and looking to obtain a more economic process. Despite being known as a pressurization process for melting alloys, die-casting can be used as a post-re-crystallisation, or a semisolid die-casting process. As a result, in this work a high pressure die casting (HPDC) system was integrated with a mechanical stir casting (MSC) system. This MSC&HPDC equipment produces components with near-net shapes in a continuous process that avoids semisolid ingot storage. It is useful for research and production of MMCs for functional and/or automotive applications. Emphasizing the need for such processes, the current automotive industry regulations for lowering emissions of CO_2 demand a significant vehicle weight reduction. Due to the recent CAFE (CAFE: Corporate Average Fuel Economy) regulations set in North America, the automotive OEMs will have to develop advanced materials and new technologies to meet the new targets set for the industry by 2025. Recent research[1] shows that approximately 50% of the powertrain components of a vehicle will have to be replaced by new parts developed with advanced materials and new technologies.

This chapter traces the development of new materials and processes with a view to improve the quality of aluminum-made parts. Since casting parts have different defectology types, (primarily porosity), net-shape processes provide a way to reduce defects and to increase mechanical properties. Additionally, the main engineering metals and alloys for different components, including aluminum alloys, now have roughly the same $E/\rho \approx 26MJ$ kg−1. Thus the only practical way to exceed this limit in a metallic material is to replace a significant fraction of the metal atoms with a new phase as happens in Al-MMC. In this work the ultimate tensile strength and yield strength at room temperature of Al-Si7-Mg0.3-T6/βSiC-15wt% composite manufactured by MSC were increased by 73% and 92%, respectively, compared to those of the original alloy with no reinforcement. The elongation of the composite material was decreased 44% because the reinforcement effect.

2. Materials

MMC are mostly aluminium-based alloys reinforced with particles. These alloys include pure aluminium, high-resistance alloys and the very common foundry Al-Si alloys, which allow the synthesis of light composites. In our case it is also important to consider an adequate gap temperature in the semisolid state of the alloy. Thus an Al-Si alloy has been used. Table 1 lists its chemical composition obtained using a Shimadzu spectroscopy 5500-OES.

Si	Fe	Cu	Mg	Mn	Zn	Al
6.5 -7.5	0.2 max	0.2 max	0.25-0.45	0.1 max	0.1 max	balance

Table 1. Chemical composition of the alloy A-356 (wt%).

The mechanical properties of ASTM test specimens made from MMC typically match or approach many of the characteristics of iron castings and steel, at lighter weight. Properties can exceed those of most Al, Mg, Zn or Cu components commonly produced by die casting. Aluminum MMC parts offer higher stiffness and thermal conductivity, improved wear resistance, lower coefficient of thermal expansion, reduced porosity, and higher tensile and fatigue strengths at elevated temperature, with densities within 5% of Al die casting alloys. In addition, particles used as reinforcement are generally less expensive than other reinforcement materials, such as fibers, because of their abundance. Also, some ceramics have far better properties in finely divided form. Notably,

---

[1] See: http://www.greencarcongress.com/ (August 2011)

micrometer-sized ceramic fibers or ceramic particles can be much stronger than bulk ceramics. Additionally, small-single-crystal ceramic particles can be excellent conductors of heat. There is a large variety of reinforcing ceramic materials. MMCs commonly are reinforced with silicon carbide (SiC) particles, because of SiC's excellent physical and mechanical properties. SiC can be obtained by several mechanisms, one of which is of particular interest as it contributes greatly to minimizing waste; this procedure involves the controlled pyrolysis of rice husks (RH).

The production method of Martínez & Valencia (2003), designed for the synthesis of SiC from RH, starts with the cleaning, sizing and conditioning the RH before the pyrolysis process. This conditioning is made up of the removal of garbage, size classification by sieving and the use of catalysts to increase the efficiency of the process. $FeCl_2.4H_2O$ was used as a catalyst and NH_4OH is used as an agent for precipitation of Fe. Pyrolysis is achieved through controlled thermal decomposition of the RH at 1370°C in an argon atmosphere for 40min. The final product is ground and subjected to a pneumatic separation process and characterization.

Fig. 1 summarizes the analysis of the resulting SiC particles via SEM. Semi-quantitative analysis by EDS yielded the following values: 68.99 wt%C, 23.99 wt%O, 6.42 wt%Si and 0.59 wt%Fe, the latter as a result of the catalyst.

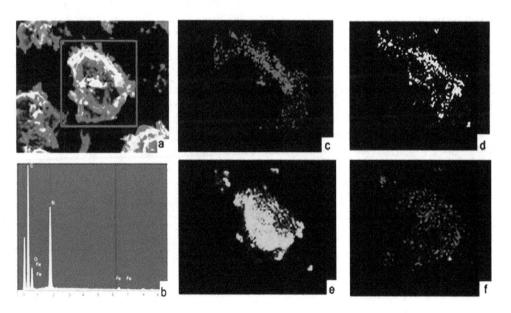

Fig. 1. Elemental mapping of SiC particles obtained by pyrolysis of RH: (a) SEM image, (b) EDS trace (c) C, (d) O, (e) Si, (f) Fe.

The XRD spectrum from the resulting sample has strong peaks at 41.58° and 48.40°, confirming the formation of SiC crystals of the Moissanite (β-SiC) variety (Fig. 2).

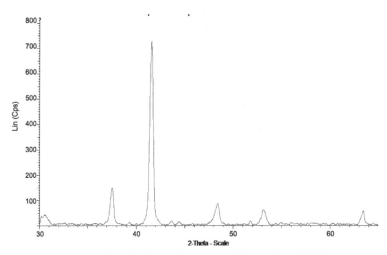

Fig. 2. XRD spectrum of pyrolyzed RH, confirming the formation of SiC.

3. Experimental procedure

In order to synthesize the composite material the difficulties associated with ceramic-metal incompatibility must be resolved; these difficulties are described in terms of the wettability between SiC and aluminum. In addition, avoiding air engulfment during the immersion and dispersion of the reinforcement particles must be ensured. These two requirements and other details of the synthesis process are discussed below.

For semisolid casting, this synthesis process is started with a partially molten aluminum matrix. Once the proper dispersion for the phases has been attained, the conformation process takes place. In this stage the fluid is translated into the cavities of the die casting device, finishing with heat treatments to ensure the required mechanical and physical properties are realized (Fig. 3).

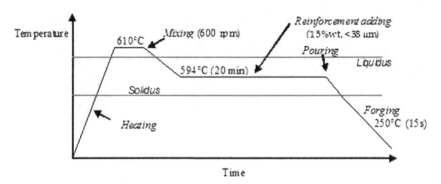

Fig. 3. Production method for Al-MMC.

3.1 Ceramic-metal compatibility

The wettability of SiC by aluminum presents two main difficulties. First, the liquid aluminum reacts with the SiC to form Al_4C_3 near the interface through a dissolution-precipitation mechanism. The formation of Al_4C_3 can be controlled by using low temperatures (<700°C), which also avoid the sublimation of Mg (vital in the subsequent heat treatment), as well as by minimizing the time between mixing and pouring. In our case, the estimated time was less than 50min. The second difficulty is that the liquid aluminum is normally covered by an oxide layer that inhibits wetting (Laurent, V. et al., 1996). Several possibilities are available to improve wetting between SiC and Al, such as inclusion of reactive elements (e.g. Li, Mg, Si) (Martínez & Valencia, 2004) or using metallic coatings to generate a metal-metal interface, and hence a higher wettability. Previous investigations have shown that the use of Ni or Cu coatings is effective, resulting in an increase in the resistance of the composite, its toughness and a better dispersion of the reinforcement particles within the molten metal (Ghomashchi, 2000). To solve the wetting problems, from the experimental model proposed by Sharma et al. (2006), we proceeded to cover the SiC with Cu deposited by electroless plating (EP); the specific pretreatment, coating and drying, processes are explained in the following section.

3.1.1 Pretreatment

Since the EP technique is a chemical reduction process, the preparation of the surface where the metal will be deposited is essential. That is why the particles were immersed in $SnCl_2$ and $PdCl_2$ solution, respectively. Table 2 lists the details of the pretreatment of the particles.

Solution	Chemical Species	Concetration	Operation	Time (min)
Sensitization	$SnCl_2 \cdot 2H_2O$ HCl	20g/l 0.5 ml/l	Mechanical Stirring (400 rpm)	30
Water wash (pH 7.0) and vacuum filtration				5
Activation (2000 ml)	HCl $PdCl_2$	5.5ml/l 0.25g/l	Mechanical Stirring (400 rpm)	35
Water wash (pH 7.0) and vacuum filtration				5

Table 2. Superficial Pretreatment of β-SiC.

3.1.2 Coating and drying

Once the particles were catalyzed, they were taken to the plating bath, which is constantly stirred, in which the reduction reaction took place (Fig. 4). This bath consists of a metallic ions solution of cupric sulfate, formaldehyde as a reducing agent and sodium-potassium tartrate (Rochelle's salt) as a complexing agent, which keeps the metallic salt from precipitating. Stirring was maintained and, as the reaction ran out, the solution color changed until it becomes transparent (Fig. 4c), which corresponded to when the Cu from the solution was deposited on the surface of the ceramic particles as metallic Cu.

The temperature of the bath was kept at 80°C while the pH was kept at 12.0. Table 3 summarizes the details of the plating process. As soon as the coating reaction was completed, the particles were washed with water and then dried in a vacuum (~1 bar) at 60°C for one hour.

Fig. 4. Evolution of the plating bath, (a) 0 min, (b) 10 min and (c) 40 min.

Solution	Chemical Species	Value	Role in the bath	Operation	Time (min)
Coating (3000ml)	CuSO₄5H₂O	10g/L	Metal ions Coating	Mechanical stirring (1200rpm)	To complete reaction
	CH₄O₆NaK4H₂O	50g/L	Complexing		
	HCHO	15m/L	Reducer		
	NaOH	To adjust pH	Buffer solution for pH control		
Water wash (pH 7.0) and vacuum filtration					5

Table 3. Parameters for the electroless plating of β-SiC.

Fig. 5a is a SEM image of β-SiC particles after the sensitization bath. Fig. 5b shows the qualitative analysis, indicating the presence of Sn, which was used as a catalytic material in the SnCl₂ bath. Fig. 5c shows the coated β-SiC particles. Again, elemental analysis shows the presence of Sn and Pd from the pretreatment process. A high amount of Cu can be observed due to the plating process (Fig. 5d).

By modifying the metal-ceramic interface (SiC-Al) by with a metal-metal type interface (Cu-Al), the micro-composites Cu/β-SiC developed to this point can be introduced into the aluminum to the synthesis of Al-MMC/β-SiC. Kim & Lee (2005) have shown that the sintering of Al-MMC/10wt% SiC, after SiC coating with 8wt%Cu, is significantly improved, achieving a further increase in bending strength. However, the addition of Cu to SiC, with a view to the synthesis of particulate composites Al/β-SiC is limited by the formation of inter-metallic compound CuAl₂ of fragile nature. The formation of this compound is subject to the solubility of Cu in Al, which according to Kim & Lee (2005) is up to 2wt%. This means that contents at or below 2wt%Cu in Cu/β-SiC avoid precipitation of Cu and the generation of unwanted inter-metallic compounds. The amount of Cu deposited is a function of coating thickness. In this case it has acted to control the thickness to approximately <0.6μm.

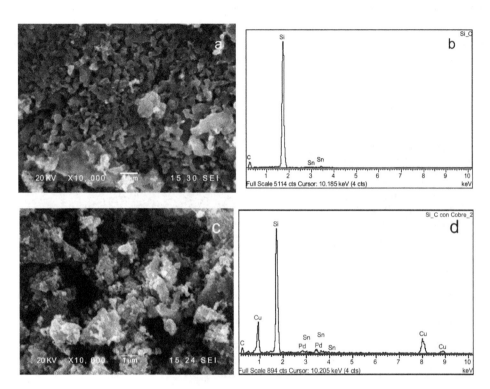

Fig. 5. SEM and EDS for (a), (b) SiC particles without Cu; (c), (d) SiC particles modified with Cu.

3.2 Mechanical Stir Casting (MSC)

In the case of aluminum-silicon alloys, the microstructure is by nature a dendritic type (Yang et al., 2005), which is commonly known as a morphology which decreases the strength of the material depending on the spacing of secondary dendritic arms. During the stirring action it was sought to convert the aluminum microstructure from the semisolid dendritic to a globular form (Mada & Ajersch, 1996). The technique is to generate shear stresses of sufficient magnitude by means of mechanical agitation, causing relative movement between layers and interlaminar friction, that in the best cases, generates a globular morphology (pseudo spheres of about 60 to 72 microns) (Fan, 2002). Nevertheless, a high shear index can be the source of the engulfment of impurities, increasing power consumption and the stress in the rotor system (Biswas et al., 2002). This is why the geometry of the stirring system is a key part in the process. The efficiency, the quasi-isotropy of the material, the micro-structural changes and the transformation the thixotropic matrix depend on this. Similarly, the agitation system must allow the distribution of reinforcement in the matrix to be uniform, depending on the stirring speed and time in steady state for the molten metal.

With the purpose of increasing the shear efficiency, preventing the formation of an external vortex and obtaining a proper dispersion of the reinforcing material, two trowel systems were studied. In both systems the mechanism was optimized with a set of trowels at 90°-y in the bottom to avoid the sedimentation of non processed material. The two configurations of the upper set of trowels are shown in Fig. 6.

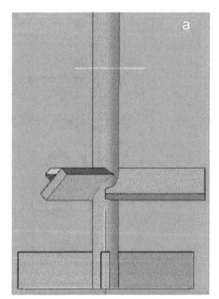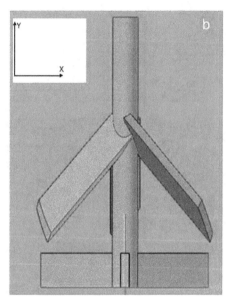

Fig. 6. Two configurations of upper trowels: (a) 45°-x, (b) 45°-x/45°-y.

To ensure adequate mixing conditions the most effective approach is the simulation of the mechanical stir casting (MSC). In this sense, simulations of flow (Flow-3D ®) were initially carried out using computational fluid dynamics (CFD) for each type of agitator to identify the shear rates and changes in material viscosity. The actual process parameters are included in the parameterization of the simulation software (Fig. 3).

3.2.1 Strain rate

The 45°-x/45°-y agitator showed a higher level of shear stress in the fluid, and greater turbulence compared with the 45°-x agitator (Fig. 7a, 7b). For the needs of the mixing process the 45°-x/45°-y stirrer is more efficient as it creates greater strain in the material at the same engine speed. By achieving greater strain on semisolid material a lower viscosity was obtained, ensuring the optimum rheological conditions for the dispersion of the particles.

3.2.2 Viscosity

In both cases it is apparent that the viscosity increases in areas where the agitation generates a low speed level in the fluid, which turns out to be useful to generate tixtotropic behavior in the fluid. The $45°$-x/$45°$-y agitator (Fig. 7d) shows a higher viscosity than the $45°$-x (Fig. 7c).

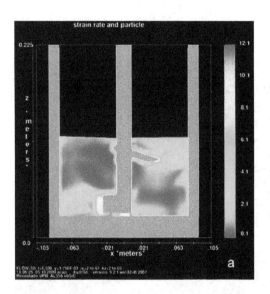
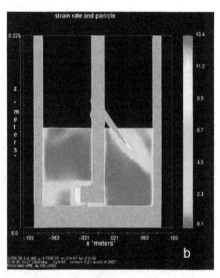
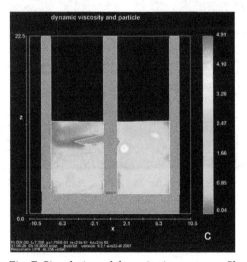
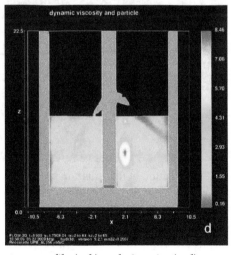

Fig. 7. Simulation of the agitation process. Shear stress profile (a, b) and viscosity (c, d). Agitator $45°$-x; (a, c), agitator $45°$-x/$45°$-y (b,d).

3.2.3 Velocity magnitude

The analysis of the agitation a few seconds after the start of the process showed that the fluid velocity reaches a higher value with the agitator 45°-x/45°-y, approximately 26 m/s, while the agitator 45°-x reaches 0.13 m/s (Fig. 8a, 8c).

The dispersion of particles with the 45°-x/45°-y agitator, when it is not completely immersed, is not as effective as with the 45°-x agitator, and the result is that the particles flow to the bottom of the crucible faster (Fig. 8c). However, the 45°-x and 45°-x/45°-y configurations allow a homogeneous particle distribution in the radial direction of the crucible (Fig. 8b, 8d).

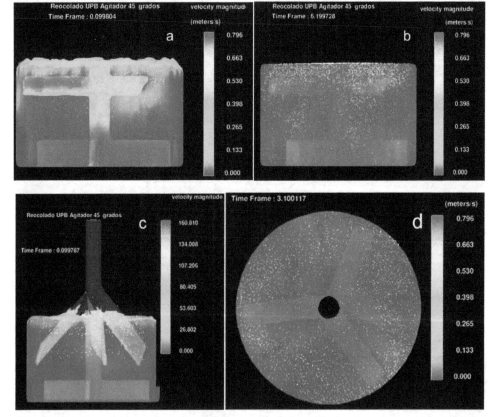

Fig. 8. Simulation of the agitation process. Stirrer speed profile for the 45°-x agitator (a, b), and for the 45°-x/45°-y agitator (c, d).

3.2.4 Air engulfment

Finally, according to the simulations the condition of air trapping in both mixers is minimal. Neither case produces a pronounced vortex that allows the entry of air into the molten aluminum. At the start of agitation during the first 6s, the trapped air was ~14vol% (Fig. 9a,

9c). However, after stirring for 20s, the fraction of trapped air drops to ~0.4vol%, indicating that the agitation is effective and does not introduce air into the melt (Fig. 9b, 9d).

For simulations of the MSC process, the 45°-x/45°-y agitator was selected, since it produced the best results in productivity and process efficiency. Then we proceeded to the synthesis of composite material. In order to control the MSC process in real time (rpm, rotation direction and geometric localization of the stirrer), a LabView® interface was implemented using a National Instruments USB data acquisition (DAQ) device.

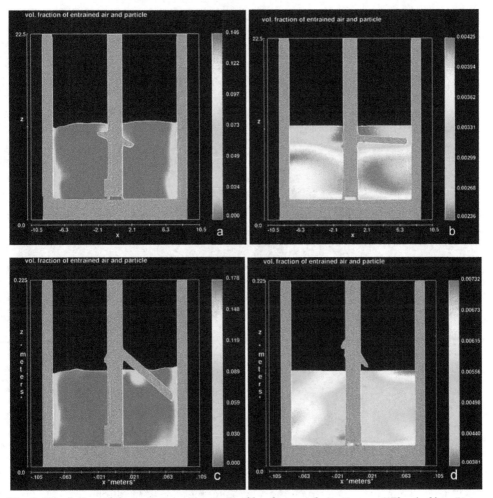

Fig. 9. Simulation of the agitation process. Profile of trapped air stirrer at 45°-x (a, b), agitator 45°-x/45°-y (c, d). stirring during first 6s (a, c), after 20s (b, d).

Table 4 lists the key parameters used during MSC. The solid fraction is 0.4. Stirring is performed in an argon atmosphere and in an anti-clockwise direction.

Parameters for MSC	Value
Melting temperature (°C)	610
Process temperature (°C)	594
Velocity magnitude (rpm)	600
Stirring conditions (min, °C)	20, 594
Reinforcement fraction (wt%)	15
Reinforcement size (μm)	<38

Table 4. General parameters for MSC.

Fig. 10 illustrates the microstructural evolution of the semisolid processed material. The resulting morphology consists of α globular phases with a diameter of about 70μm, surrounded by eutectic microconstituents.

Fig. 10. Optical images of microstructure evolution for the alloy A-356 during the MSC. (a) dendritic structure (50X), (b) rosette type morphology (50X), (c) formation of globular structure (50X), (d) globular structure (including reinforcement particles 50X).

Figs. 11 and 12 shows the morphology obtained after the addition of reinforcement material. Fig. 11a shows the alloy processed by MSC with β-SiC particles. The morphology of the matrix is completely globular, with grain sizes ranging from 75 to 100µm. In Fig. 11b, Fig. 12a and 12c the reinforcement particles are fully dispersed and preferentially located in the eutectic zone. Some faceted forms for the reinforcement particles were observed (Figs. 11b, 12c), which creates fewer opportunities for a mechanical interface with the metal matrix. It can be argued that the interface must be chemical, as it was envisaged during the EP treatment of the reinforced particles, modifying the metal-ceramic interface (SiC-Al) into a metal-metal type interface. At the experimental level there is minor porosity, which must be corrected in the subsequent HPDC process.

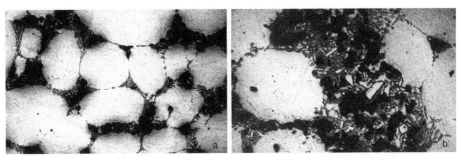

Fig. 11. Optical images of the microstructure of A-356/βSiC-15wt% composite at (a) 50X and (b) 100X.

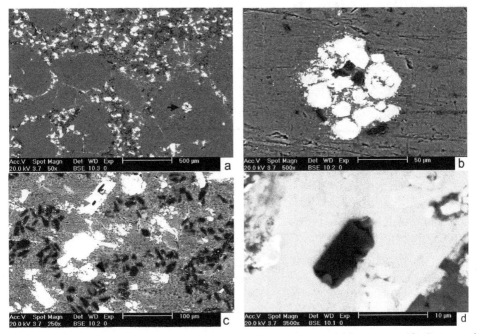

Fig. 12. SEM images of the A-356/βSiC-15wt% composite; (b) and (d) are magnified views of (a) and (c), respectively, showing reinforcement particles engulfed in eutectic zones.

3.3 High Pressure Die Casting (HPDC)

After obtaining a homogeneous composite, a die cast is made in order to obtain an ingot by applying moderate pressure for 12 to 15s. Our MSC&HPDC (Fig. 13) is useful both for research and production on a laboratory scale. The HPDC has four hydraulic cylinders to provide the load. The HPDC process for an ingot of semisolid material and its solidification take place when a pressure ranging from 50 to 100 MPa is applied. Other parameters are listed in table 5. This device has no striker ejection pin; instead, after the hydraulic cylinders are opened, the part falls to a container located at the bottom of the device.

Parameters for HPDC	Value
Pouring temperature (°C)	594
Mold temperature (°C)	250
Load (MPa)	50-100

Table 5. General parameters for HPDC.

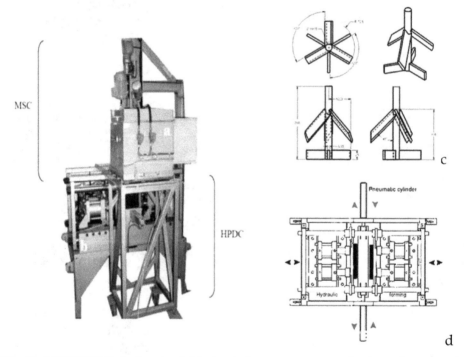

Fig. 13. MSC&HPDC device: (a, c) Mechanical stir casting unit; (b, d) High pressure die casting (shaping) unit.

3.3.1 Computational fluid dynamics simulation

The semisolid forging process of a piece was initially simulated using CFD tools (Flow-3D®). Fig. 14 shows an image of the simulated part. With this simulation it is possible to determine zones where defects could develop instantly, as well as sites where interlaminar differences could generate irregular and turbulent flow. This figure also shows the velocity profile in three different moments for the piece. It can be seen that in the external part of the mold there is an increase in the velocity of the flow. Despite the high velocity, at each of the time intervals analyzed, the outer or leading surface of the material inside the mold is homogenous, thus avoiding defects by gas inclusion.

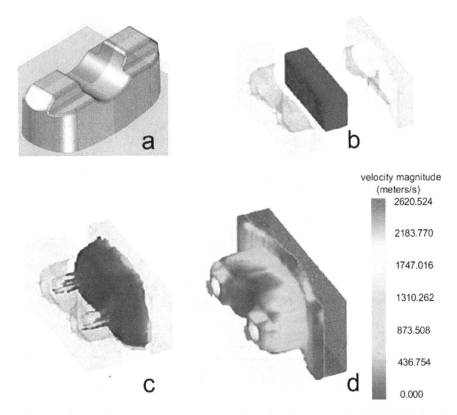

Fig. 14. Computational fluid dynamics analysis of piece to be created by HPDC (a) shape of the final part , (b) initial position of the ingot and mold, (c-d) velocity profiles at two stages of pressing of the composite paste.

3.3.2 Manufacture and heat treatment

After verifying the integrity of the piece through the simulation of the HPDC process, manufacture was carried out using the parameters of pressure and application load times mentioned above. Importantly, prior to the HPDC process the preformed mold and the mold (Fig. 14b) were heated to 300°C. The piece underwent thermal treatment in solution at 548°C for 8h, followed by cooling in water and artificial aging (T6) at 170°C, for 6h and cooled with air. Fig. 15 shows the HPDC piece and molds.

__0,5cm

a   b

Fig. 15. (a) HPDC molds, and (b) resultant piece produced by MSC&HPDC.

4. Mechanical characterization

The dispersion of stiff ceramic particles or fibers within a ductile metallic matrix leads to an increase in flow stress of the metal by load transfer across a strong interface from the matrix to the reinforcement (Mortensen & Llorca, 2010). Table 6 summarizes the mechanical properties of the materials processed by MSC&HPDC as compared with nominal A-356 T6 alloy with no reinforcement (Fig. 16a). It is evident that an increase in mechanical resistance and a reduction in the elongation percentage were achieved, demonstrated by the high rigidity of the composite compared with the nominal alloy.

Property	Al-Si7-Mg0.3-T6	Al-Si7-Mg0.3-T6/β-SiC-15wt%
Ultimate strength (MPa)	220	380
Yield strength (MPa)	180	345
Elongation percentage (%)	18	10
Hardness (HB)	110	130

Table 6. Mechanical properties of processed materials

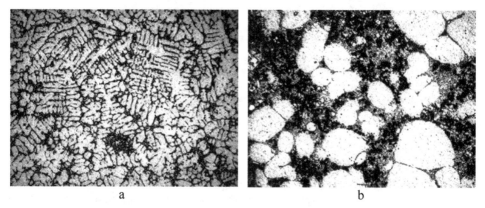

a   b

Fig. 16. Comparison of microstructures of materials processed by MSC&HPDC at 50X (a) Al-Si7-Mg0.3-T6 alloy and (b) (Al-Si7-Mg0.3-T6/β-SiC-15wt%)-T6 composite. Both parts were thermally treated in solution at 548°C for 8h, followed by cooling in water and artificial aging (T6) at 170°C, for 6h and cooled with air.

5. Conclusions

Although research on the subject of MMCs reached a high level of intensity in the late 1980s and early 1990s, interest continues today, albeit in a wider array of more distinct directions, as for example, reported here for the production of pieces from Al-MMC/β-SiC that can be used in different applications.

Micrometric β-SiC particles, obtained by controlled pyrolysis of rice husk have been modified with Cu through electroless plating (EP). EP coatings enhance the adhesion between Al and β-SiC.

A novel technique for manufacturing parts from MMCs has been designed and tuned. Since the device and processing routes are similar to the conventional ones used in casting, it is safe to say that mechanical stirring coupled with semisolid forging shows interesting advantages, including its low cost, easy process implementation, and the ability to form diverse, near-net-shape parts.

A composite Al-Si7-Mg0.3-T6/β-SiC-15wt% has been processed by MSC&HPDC and improved mechanical properties were achieved. This process could be useful for producing parts for various industries, including automotive applications.

6. Acknowledgment

This work has been made possible thanks to the support of COLCIENCIAS, the Engineering School of Antioquia and the Pontificia Bolivariana University in Medellin. Additional thanks to Dr. Mike Boldrick for his valuable suggestions on this document.

7. References

Biswas, P., Godiwalla, K., Sanyal, D., & Dev, S. (2002). A simple technique for measurement of apparent viscosity of slurries: sand–water system. *Materials and Design*, Vol. 23, No. 5, pp. 511-519. ISSN 0261-3069.

Fan, Z. (2002). Semisolid metal processing, *International Materials Reviews*, Vol. 47, No. 2, (April 2002), pp. 49-85, ISSN 0950-6608.

Ghomashchi, M. R. & Vikhrov, A. (2000). Squeeze casting: an overview. *Journal of Materials Processing Technology*. Vol. 101, (April 2000), pp. 1-9. ISSN 0924-0136.

Kim, I. S. and Lee, S. K. (2005). Fabrication of carbon nanofiber/Cu composite powder by electroless plating and microstructural evolution during thermal exposure. *Scripta Materialia*, Vol. 52, No. 10, (May 2005), pp. 1045-1049. ISSN 1359-6462.

Mada, M., & Ajersch, F. (1996). Rheological model of semisolid A356-SiC composite alloys. Part I: Dissociation of agglomerate structures during shear. *Materials Science and Engineering A*, Vol. 212, No. 1, (July 1996), pp. 157-170. ISSN 0921-5093.

Martinez, H. V.; Valencia M. F.; Chejne, F. & Cruz, L. (2006). Production of β-SiC by pyrolysis of rice husk in gas furnaces. *Ceramics International*, Vol. 32, (December 2006), pp.891–897, ISSN 0272-8842.

Martinez, V.; Valencia, M. Compocasting process of cast Zamak particulate composites. Frontiers in Materials Research, A *CIAM-CIMAT-CONICYT WORKSHOP*, p. 68, Viña del Mar, Chile, 26-29 April 2004.

Mortensen, A., & Llorca, J. (2010). Metal Matrix Composites. *Annual Review of Materials Research*, Vol. 40, No. 1, pp. 243-270.

Sharmaa, R., Agarwala, R.C. & Agarwala V. (2006). Development of copper coatings on ceramic powder by electroless technique. *Applied Surface Science*. Vol. 252, No. 24, (October 2006), p.p. 8487-8493. ISSN 0169-4332.

V. Laurent, C. Rado & N. Eustathopoulos (1996). Wetting kinetics and bonding of Al and Al alloys on α-SiC. *Materials Science and Engineering A*. Vol. 205, (January 1996), pp 1-8, ISSN 0921-5093.

Yang, Z. Kang, C.G. & Seo, P.K. (2005). Evolution of the rheocasting structure of A356 alloy investigated by large-scale crystal orientation observation, *Scripta Materialia*, Vol. 52, No. 4, (February 2005), pp. 283–288, ISSN 1359-6462.

7

Squeeze Casting of Al-Si Alloys

Bellisario Denise[1], Boschetto Alberto[2], Costanza Girolamo[1],
Tata Maria Elisa[1], Quadrini Fabrizio[1] and Santo Loredana[1]
[1]*Department of Mechanical Engineering, University of Rome Tor Vergata,*
[2]*Department of Mechanics and Aeronautics, University of Rome La Sapienza,*
Rome,
Italy

1. Introduction

Nowadays there is a great demand of lightweight parts with high mechanical performances; in such applications, e.g. automotive, production rate and process control are critical aspects as well. This way, pressure assisted casting processes are a good compromise between performances and costs. Over the past decade, aluminium or magnesium components produced by means of pressure assisted processes have been introduced to substitute cast iron components. When the substitution is possible, a great advantage in terms of cost saving and component weight reduction is obtained.

The main technological issue in the development of new pressure assisted casting processes is related to the increase of the holding pressure during the alloy solidification. In fact, it is evident that, by increasing the pressure, the overall quality of the casting generally increases in terms of a smoother surface, lower porosity and higher mechanical properties. In die casting, higher holding pressures lead to higher production runs too. However, it is not yet clear what is the main role of the pressure during casting as many other process parameters are present and their effect cannot be easily separated from the effect of the pressure. Moreover, different results may be obtained depending on the alloy to cast and on its sensibility to the casting process. In this study, several experiences were collected which show the effect of the pressure during the squeeze casting of Al-Si alloys. The squeezing systems were designed on purpose to point out different aspects of the squeezing process: the effect of the pressure at constant cooling rate, the mechanical property distribution in the casting, the expected properties for the Al-Si alloy. Numerical modelling of the squeezing operations allowed to evaluate the limits of the production systems and to provide general conclusions. The combination of the proposed squeezing systems and the related numerical models is the main result of the current work as well as the prediction of the Al-Si alloy mechanical properties as a function of the squeezing pressure.

For aluminium alloys, the application of a holding pressure during cooling plays a very important role in defining casting properties. A significant microstructural refinement was already observed during the solidification of an Al-Cu alloy under high pressure (Han et al., 1994). In this case the average dendrite cell size changes from 30 μm, without applied pressure, to 5 µm under 1.7 GPa. For an Al-Si alloy (designated as B390) a great microstructure refinement was also observed as a consequence of the pressure application (up to 100 MPa) and higher mechanical properties were measured in terms of hardness and tensile strength (Maen et al., 2000; Lee et al., 2000) but no material model has never been proposed. Other studies showed that in direct and indirect squeeze casting, apart from the pressure, other process parameters strongly affects microstructure and mechanical properties such as melt and die temperature (Maen et al., 2000; Lee et al., 2000), die geometry (Kim et al., 1999) and melt flow (Lee et al., 2000). In such conditions, it is very difficult to understand what process parameter is the main one in determining the final mechanical properties of the cast alloy. As pressure increases, together with microstructure refinement, other structural modifications can occur such as shrinkage, porosity change (Hong et al., 1998a) and macrosegregation (Hong et al., 1998b; Gallerneault et al., 1995; Gallerneault et al., 1996). As a result, a complete process map can be obtained with the definition of the process window where no micro and macro-defects occur (Hong et al., 1998a; Hong et al., 1998b). Also for aluminium matrix composites, a positive effect of the applied pressure was observed for mechanical properties (Shuangjie and Renjie, 1999). However all these studies do not show a way to predict the final performances of the cast alloy depending on the casting conditions.

From a theoretical point of view, the effect of the pressure application can be divided into two contributions: a direct and an indirect one. The first one is related to the alloy phase diagram modification. Thermodynamic equilibrium curves shift toward higher temperatures as pressure increases. For aluminium alloys (and generally for metals) the transformation temperature shift is expected to be negligible as the phase diagram of the alloy minimally changes with the pressure (Savas et al., 1997). A second direct effect is related to the porosity reduction. In fact at high pressures, entrapped gases remain in solution without cavity formation (Kalpakjian, 2000). Also shrinkage porosity formation is prevented. As a consequence, the resulting metal density is higher. A last direct effect is also macrosegregation (Hong et al., 1998a) even if a strong dependence on the cooling rate is also present. On the other hand the pressure indirect effect is related with the cooling rate. Allowing the better contact between the mould and the cast part during solidification, pressure enhances heat transfer, increasing cooling rate. As a consequence, a lower grain size is observed inside the component also far from the external surface. Actually, pressure is not the only parameter which influences the cooling rate (the melt and die temperature, the melt flow condition and the process times are important as well) but it's surely the most easily adjustable.

At present, it's not clear if direct and indirect pressure effects have the same importance in defining the cast mechanical properties. This way, it's not possible to establish if a pressure increase could be a convenient practice when porosity is not a problem and significant changes of the cooling rate are not expected. It is not a secondary problem as a pressure increase leads to a cost increase. On the other hand, the cooling rate evaluation is generally a very difficult task for all the casting processes and in particular for aluminium alloy squeeze casting. In fact, cooling rates can reach very high values and a direct temperature measurement is practically impossible due to the mould thermal inertia. Cooling rate is never directly measured except for some cases when thermocouples are inserted inside the casting. Lee et al. studied the effect of the gap distance on the cooling behaviour and the microstructure of an indirect squeeze cast and gravity die cast aluminium alloy. They inserted thermocouples through the die, located at the centre of each casting. However a numerical model was used to calculate the temperature profile and was preferred for the cooling rate extraction due to the intrinsic dispersion of the measured cooling curves (Lee et al., 2002). In fact, too many factors affect the temperature profile during casting, from melt turbulence during injection to mould transient phase. Britnell and Neailey placed thermocouples inside the die near the mould-melt interface to study the macrosegregation in thin walled squeeze cast samples but thermal traces can be used just for a qualitative analysis (Britnell et al., 2003).

In order to study squeeze casting, it is very important to evaluate the influence of the pressure direct and indirect effects. For this purpose, in the current study, two sets of Al alloy samples were fabricated: the first set by using different applied pressures and same cooling rate; the second one by using the same applied pressure but different cooling rates. Moreover, a finite element (FE) model was set up so as to simulate the thermal history inside the mould during the heating-cooling sequence. A maximum cooling rate was estimated in this case, resulting from the chosen mould geometry.

Another important aspect under investigation is the material behaviour simulation and the prediction of mechanical property distribution in large ingots. A different squeezing system was designed for this purpose and other two sets of samples were produced. A suitable FE model was also defined which permits to gather the properties of an aluminium part when its geometry and process history are known. In a reverse approach, as local mechanical testing can be always performed on cast parts, the cooling rate could be inferred from the mechanical properties resulting from the tests. This information can be used for process optimization as well as for the validation of process numerical simulations.

2. Material modelling

A simple law which correlates mechanical properties (yield strength and Vickers hardness) with cooling can be extracted from the scientific literature. First, the average grain size can be correlated to the mechanical properties by Hall-Petch (Reed Hill, 1996):

$$\sigma_Y = \sigma_0 + K_Y \lambda^{-\frac{1}{2}} \tag{1}$$

$$HV = HV_0 + K_H \lambda^{-\frac{1}{2}} \tag{2}$$

This equation was also proposed for the prediction of mechanical properties of reinforced aluminium specimens (Shuangjie and Renjie, 1999). Furthermore, a simple relationship between the cooling rate (ε) and the average dendrite cell size (λ) was used for aluminium alloys too (Han et al., 1994; Kim et al., 1999):

$$\lambda = B\varepsilon^n \tag{3}$$

Combining equations (1), (2) and (3), a direct correlation between the cooling rate and both the final yield strength and the hardness is obtained:

$$\sigma_Y = \sigma_0 + C_Y \varepsilon^m \quad \text{where} \quad \begin{cases} C_Y = K_Y B^{-\frac{1}{2}} \\ m = -\dfrac{n}{2} \end{cases} \tag{4}$$

$$HV = HV_0 + C_H \varepsilon^m \quad \text{where} \quad C_H = K_H B^{-\frac{1}{2}} \tag{5}$$

If the pressure indirect effect is predominant on the direct one, equations (4)-(5) can be used in order to predict the alloy mechanical properties with pressure during solidification.

3. Aluminium alloy

The examined material, aluminium alloy EN-AB46000, has nominal composition (wt.%) showed in the following Table 1. EN-AB46000 is a hypoeutectic Al–Si alloy, normally used in die casting processes. This alloy is universally known for its good castability, except for a little tendency towards forming surface and internal cavities caused by shrinkage during solidification.

Si	Fe	Cu	Mn	Mg	Cr	Ni	Zn	Pb	Sn	Ti	Al
8.0–11.0	0.6–1.1	2.0–4.0	0.55	0.15–0.55	0.15	0.55	1.20	0.35	0.25	0.20	To balance

Table 1. Al alloy composition (EN-AB46000).

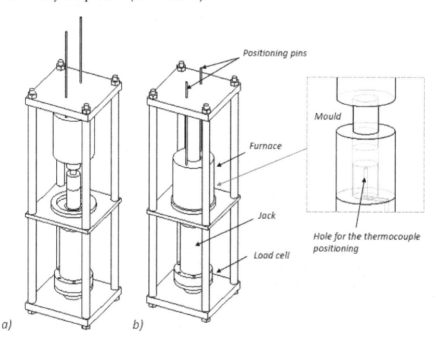

Fig. 1. 3D- sketch of the experimental apparatus: (a) Furnace position during the solidification; (b) Furnace position during the melting.

4. Study of the effect of the squeezing pressure

4.1 A squeezing system for small samples

In order to underline the effects of the pressure in squeeze casting, an original experimental apparatus was built. This kind of apparatus was made to cast samples under different pressure and cooling rate conditions. The experimental system was designed to melt the aluminium alloys directly in the mould and to apply the established pressure before starting the sample solidification. An on line acquisition system was defined for temperature and pressure data. The apparatus (Figure 1) consisted essentially in a frame, a mould, a resistance furnace, a thermocouple, a load cell and a jack. The cylindrical-shaped mould was enclosed by the coaxial resistance furnace (power 700 W). The furnace could be moved after the sample melting. The piezoelectric load cell, located near the furnace below the jack, had the maximum load of 1000 kg. Considering that the mould cavity had a 15 mm diameter, the maximum achievable pressure was about 56 MPa. The thermocouple was placed inside the mould in a hole made at the bottom. The AISI 303 steel was chosen to fabricate the frame and the mould.

During experimentation, the aluminium alloy was melted inside the mould and subsequently it was squeezed until its solidification. Two different process conditions were used. For the first condition, several pressures were applied, leaving the cooling rate approximately constant. In the second condition, the pressure was kept constant whereas the cooling rate was changed. The same initial melt temperature was considered in both cases. Due to the high thermal capacity of the steel mould in comparison with the aluminium alloy sample, the sample cooling rate depended only on the heat transfer between the mould and the air and no effect of the applied pressure was observed.

First fabrication condition					
Applied load [N]	Applied pressure [MPa]	Initial melt temperature [°C]	Melt temperature at the pressure application [°C]	Mould cooling condition	Cooling rate [°C/s]
0	0	640	600	In air	0.7
2500-6500	14-37	640	600	In air	0.9
Second fabrication condition					
Applied load [N]	Applied pressure [MPa]	Initial melt temperature [°C]	Melt temperature at the pressure application [°C]	Mould cooling condition	Cooling rate [°C/s]
6500	37	640	600	Water spray	1.9
6500	37	640	600	Compressed air flow	5.9
6500	37	640	600	Compressed air with aqueous suspension flow	10.45

Table 2. Process parameters for aluminium alloy sample fabrication.

Table 2 summaries the process variables used for the specimen fabrication. All the samples had cylindrical shape with a diameter of 15 mm and a height of 3 mm. Each specimen was preliminarily subjected to mechanical polishing and subsequently etched with 1% aqueous HF so as to reveal its microstructure. Observations were made with an optical microscope. FIMEC (Flat-top Cylinder Indenter for Mechanical Characterization) and Vickers microhardness tests were performed on the samples.

The FIMEC test (Donato et al., 1998) is based on the penetration, at constant rate, of a flat punch of small size (a diameter of 2 mm and a height of 1.5 mm). During the test, the applied load and the penetration depth are measured, therefore load vs. penetration (LP) diagrams can be recorded. The characteristics of LP curves were described in detail in previous works (Gondi et al., 1996): the limit load q_Y is reached after an initial linear stage, which is followed by a work-hardening-like stage with loads tending to a saturation value q_S. Indeed material work hardening starts much before the q_Y load during indentation. In such standardized conditions (penetration rate $\cong$ 0.1 mm/min and deformation rate in tensile test $\cong 10^{-3}$ s^{-1}), $q_Y/A \cong 3\ \sigma_Y$ and $q_S/A \cong 3\ \sigma_U$ where A is the area of the flat pin. A WC indenter of 2 mm in diameter, a 10 N pre-load, a 0.1 mm/min penetration rate and a 0.5-1 mm penetration depth were used for testing.

4.2 Experimental results

The microstructures of two solidified samples with different pressures (14 MPa and 37 MPa) at the cooling rate of 0.9 °C/s are showed in Figure 2. In both cases, the microstructures are similar and show primary aluminium dendrites, particles and eutectic constituents. In the same figure it is reported also the comparison between the resulting mechanical data. Also for other specimens, fabricated at different pressures but with the same cooling rate, comparable microstructure and mechanical data were obtained.

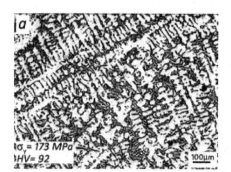 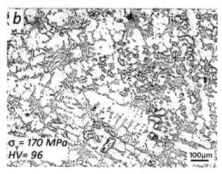

Fig. 2. Micrographs of metallographic sections and mechanical data of samples solidified at the pressure of 14 MPa (a) and 37 MPa (b) and the cooling rate of 0.9 °C/s.

Table 3 shows mechanical and microstructural data extracted from specimens produced both with the first and the second set of process parameters. In Figures 3 and 4 the micrographs of the metallographic sections of the samples are reported as well. In particular, Figure 3 shows the micrographs of an as-received-cast ingot together with micrographs of solidified samples with the first set of process parameters. In Figure 4, micrographs of samples solidified with different cooling rates but the same pressure of 37 MPa are shown.

Experimental results show that the pressure direct effect is negligible compared to the indirect one, at least in the considered pressure range. In fact, Figure 2 shows that, with the same cooling rate, no significant modifications occur in microstructural characteristics and mechanical properties. On the other hand, Figures 3-4 and Table 3 indicate that the greater the cooling rate the higher the Vickers microhardness and yield strength values due to the finer microstructure. These results were obtained thanks to the experimental apparatus, able to obtain samples at different pressures with the same cooling rate and vice versa. Moreover, the sample size is sufficiently small so as to assume that, inside the sample, an uniform distribution of the final mechanical properties is present.

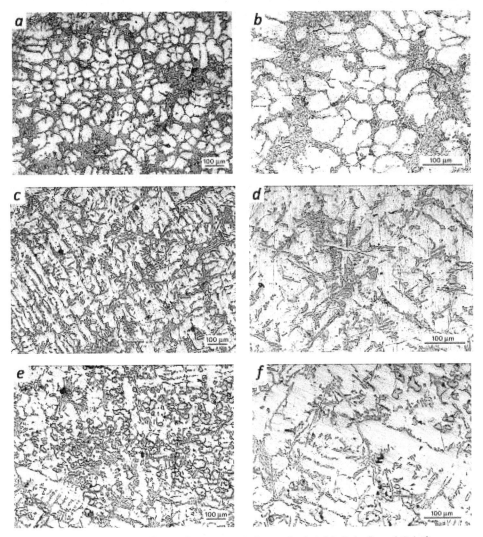

Fig. 3. Micrographs of metallographic sections of sample A (a,b), B (c,d) and C (e,f).

Some experimental thermal curves are reported in Figure 5, where the pressure effect on the cooling rate is separated from the mould cooling condition effect. The attention was focused in the cooling stage.

As a consequence of the pressure direct effect predominance, equations (4) and (5) can be used for the prediction of the material properties, considering the cooling rate as the only process parameter. Actually, the cooling rate is not directly adjustable, as it is function of the other process variables such as the melt and die temperature, the applied pressure and the cast geometry. Figures 6, 7 and 8 show that experimental data have a good fitting by means of the proposed equations (1-5). Particularly for equations (1)-(2) in Figure 6, presented data are also in good agreement with data extracted from literature (Han et al., 1994; Kim et al., 1999) for Al-Cu alloys. This occurrence suggests that the relationship between the cooling rate and the dendrite cell size could be expanded to the whole Al alloy family but further investigation would be necessary.

Some important considerations can be extracted from the experimental results so as to optimize a pressure assisted casting process. If the pressure direct effect remains negligible also for higher pressure values, there is no reason in increasing the pressure of a casting process. With the pressure level is sufficient to avoid porosity, it is advisable to use the minimum pressure value which leads to a high cooling rate. This way, it is also possible to change the other process parameters, such as the melt and die temperature, if that is a less expensive procedure. For example, in the squeeze casting process, similar mechanical properties can be obtained by reducing the pressure and decreasing contemporaneously the die temperature, if the preventive solidification can be avoided.

Samples	Pressure [MPa]	Cooling rate [°C/s]	Average dendrite cell size [μm]	HV	σ_Y [MPa]
A (as received)	0	-	48.9	81	150
B	0	0.7	37.1	92	147
C	37	0.9	36.6	94	160
D	37	1.9	33.1	95	173
E	37	5.9	20.3	117	199
F	37	10.5	16.6	116	200

Table 3. Mechanical and microstructural data extracted from the samples.

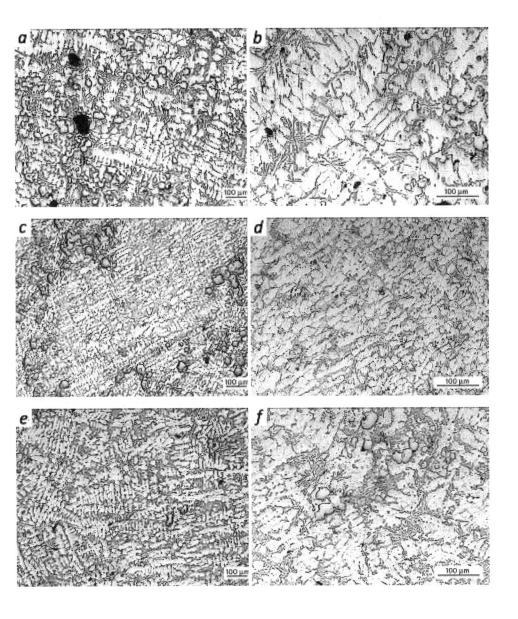

Fig. 4. Micrographs of metallographic sections of samples solidified with a pressure of 37 MPa and different cooling rates: D (a,b), E (c,d), F (e,f).

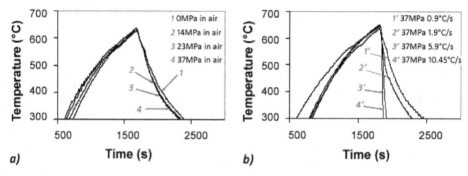

a)

b)

Fig. 5. Effect of pressure and cooling conditions on cooling rate.

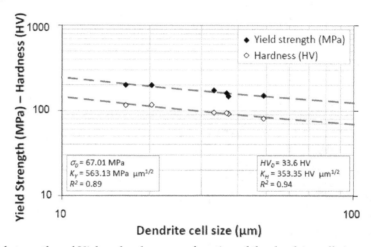

Fig. 6. Yield strength and Vickers hardness as a function of the dendrite cell size.

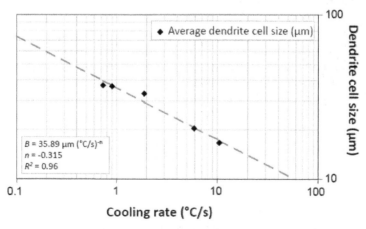

Fig. 7. Average dendrite cell size as a function of the cooling rate.

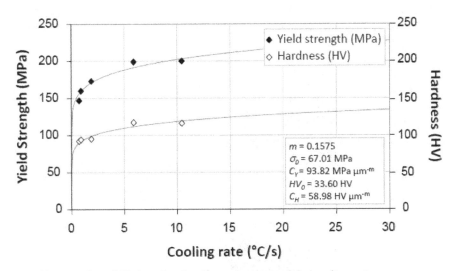

Fig. 8. Yield strength and Vickers hardness as a function of the cooling rate.

Equations (4)-(5) are particularly important, as they show the advantage that it is possible to reach by increasing the cooling rate. Evidently, it is not convenient to increase the cooling rate if the actual process is already positioned on the plateau of the curve of Figure 8. In this case, great changes of the cooling rate would produce small modifications of the mechanical properties. Equations (4)-(5) can be also used in a reverse way: measuring the properties of an aluminium alloy casting, it is possible to extract the cooling rate during its solidification. In particular, by using hardness measurements, a complete process control can be carried out if indentations are made on the overall geometry of the cast part. Dealing with the small size of the samples, the application of a miniaturized FIMEC test seems to be the only way for a proper material characterization.

4.3 Numerical simulation for the evaluation of the squeezing system performances

In order to simulate the heating-cooling sequence during the sample production, a FE axy-symmetric thermal model was built by discretising the mould geometry. The model is shown in Figure 9 where the material properties used are also reported. Generic values for an aluminium alloy and a stainless steel were used for these properties. Particularly for the aluminium alloy, a linear variation of the enthalpy was considered from the liquidus temperature to the solidus one. The applied thermal load depended on the convective heat transfer acting on the external surfaces.

Several analyses were performed by means of the described FE model. Initially the heating was modelled assuming that the convection bulk temperature T_b changes as a consequence of the furnace heating meanwhile the heat transfer coefficient h was constant and fixed arbitrarily at 50 W/(m^2 K), and the initial material temperature was 25 °C. The bulk temperature curve, as a function of time, was extracted fitting an experimental curve by means of the FE model. Being all the experimental curves identical during the heating, whatever curve could be used for this calibration. At the end the of the calibration procedure, the obtained dependence of the bulk temperature versus time (Figure 10) substitutes, inside the model, the action of the furnace which is not directly modelled.

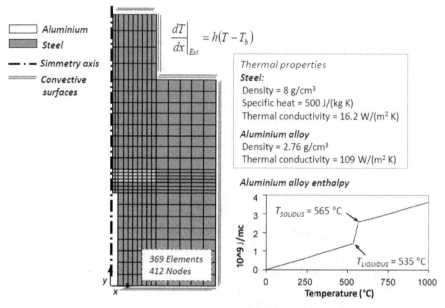

Fig. 9. The FE model.

After calibration, other thermal analyses were performed to simulate the cooling phase. In this case, at the end of the heating stage, a 25 °C constant value of the bulk temperature was applied to the convective surfaces, simulating the furnace removal. The different cooling conditions experimentally applied were simulated by changing the heat transfer coefficient h, which was always assumed constant along all the external surfaces. Figure 11 depicts the thermal map at the end of the heating phase. This map is approximately identical for all the fabrication conditions. In the same figure it is also reported the dependence of the cooling rate on the applied heat transfer coefficient. Furthermore, in the last graph, the experimental cooling rates are also reported.

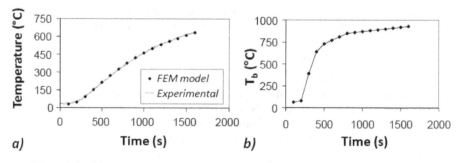

Fig. 10. FE model calibration.

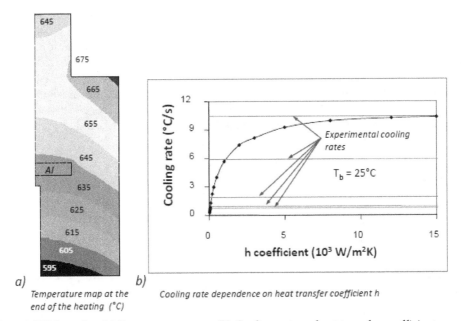

a) Temperature map at the end of the heating (°C)

b) Cooling rate dependence on heat transfer coefficient h

Fig. 11. FEM results: (a) Temperature map; (b) Cooling rate vs heat transfer coefficient.

Numerical results (Figure 11) show that a plateau in the cooling rate value is present if only the mould heat transfer condition changes. The higher experimental cooling rate is close to this numerically evaluated plateau, therefore no further significant decrease in the cooling rate can be expected by acting on the mould external walls. Actually, the proposed thermal model does not contain all the features to fully simulate the thermal problem. As an example, no thermal contact resistance was considered between the aluminium melt and the steel. This simplification is due to the fact that the temperature was experimentally acquired in only one point due to the limited size of the sample. In a more complex model, other calibrating constants would be present and more acquisition points would be necessary as well. Under these assumptions, all the neglected thermal aspects (such as the thermal resistance or the irradiation) are all taken into account globally by means of the same convective constants. However, FE modelling was important during experimentation, being possible to predict the limits of the experimental procedure and to understand the physical phenomena at the basis of the experiences.

5. Mechanical property distribution

5.1 A new squeezing system for larger ingots

In the first part of the experimentation, small samples were produced so as to increase the homogeneity of the samples; moreover the samples were directly cast in the mould to separate the effect of the pressure from the effect of the cooling rate. The second step of the study was to cast larger samples to evaluate the effect of the size on the casting properties. A laboratory squeeze casting machine was constructed on purpose to produce cylindrical aluminium alloy ingots at different squeezing pressure and mould temperature.

Two different sets of samples were produced by using the same machine but different dies; the process parameters are reported in Table 4. For both sets, the squeezing pressure was applied for 30 s and it ranged from 0 to 100 MPa. Actually, also for 0 MPa, a minimum pressure was applied to enhance the ingot solidification. The first set of samples (100 mm in length and 30 mm in radius) was produced to evaluate the effect of the squeezing pressure on the distribution of the mechanical properties. In the second set (120 mm in length and 48 mm in diameter) the mould temperature was also changed to investigate the combined effect with the pressure. Two nominal mould temperatures were used even if small deviations were measured during the experimental practice.

First set of samples					
Sample	Applied pressure [MPa]	Melt temperature [°C]	Mould temperature [°C]	Delay time [s]	Pressure time [s]
A	0	750	350	15	30
B	50				
C	75				
D	100				
Second set of samples					
Sample	Applied pressure [MPa]	Melt temperature [°C]	Mould temperature [°C]	Delay time [s]	Pressure time [s]
a	0	750	300	10	30
b	50				
c	75				
d	100				
e	0	750	350	10	30
f	50				
g	75				
h	100				

Table 4. Process parameters for the two set of samples.

In the final equipment, the lower die was designed without any mould separation (useful for the sample extraction) to avoid flash during squeezing: Figure 12 reports the squeeze casting apparatus and some cast ingots. The pressure was applied by means of the upper plug whereas the lower mould was fixed to a support that was fixed in turn on a basement. A pin was inserted in the plug to allow the ingot extraction at the end of the cooling stage. The mould presented a hole for the insertion of a K-type thermocouple. An hydraulic cylinder was used to provide the pressure on the plug that was fixed to it. A hollow thermostatic oven was used to keep in temperature the mould.

At each casting operation, a portion was cut from the as-received ingot, afterwards it was put into a graphite crucible and molten in a muffle. When the melt temperature reached a temperature about 750 °C, it was poured in the mould. The mould was preheated by means of the hollow oven. About 2 s were necessary to pour the melt into the mould and other 10-15 s elapsed before the application of the squeeze pressure (due to the time for the plug approach). After the squeezing, the hollow oven was removed and the mould was left to cool in air. During the first stages of the ingot casting, the mould temperature was acquired.

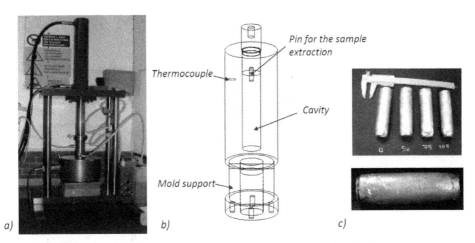

Fig. 12. (a) The experimental apparatus; (b) a sketch of the mould for the direct squeeze casting; (c) cast ingots.

At the end of the cooling, each cast ingot was extracted only by moving the plug thanks to the pin inserted into the plug (Figure 12). In order to evaluate the effect of the squeezing pressure on the ingot surface properties, the roughness of each ingot was measured along the cylinder height. Four measurements were performed at 90° of angular spacing. In analogy with the experimentation on the small samples, each ingot was cut and each section was preliminarily subjected to metallographic preparation and etched to reveal the microstructure. FIMEC and Vickers microhardness tests were performed as well. In particular, the microhardness was measured along the radius of the sections as well as the alloy dendrite size. Instead, the FIMEC test was performed only at the centre of the sections.

5.2 Experimental results

5.2.1 Effect of the squeezing phase on the mechanical property distribution

The effect of the squeezing phase on the casting properties is evident in all the ingot characteristics (Table 5). The surface aspect is clearly dependent on the pressure due to the better metal-mould matching during squeezing. Roughness is strongly dependent on pressure not only in terms of mean value over the surface but also for dispersion. Moreover, the comparison between the different microstructures, for the maximum and minimum applied pressure, in two different positions (the centre and the edge) underlines that the microstructure is strongly affected by the pressure (Figure 13). Also in these ingots, the microstructures show primary aluminium dendrites, particles and eutectic constituents but a different dendrite size distribution is observed across the sample. The dendrite size appears to be comparable for all the pressures near the sample skin.

	FIMEC Test	Roughness (Ra)			Micrographs	Microhardness
Sample	Yielding stress in the centre	Mean value	Dispersion	Distance from the centre	Dendrite cell size	Vickers microhardness
	[MPa]	[μm]	[μm]	[mm]	[μm]	[HV]
A	120.4	3.08	0.82	1	50.1	80.2
				6.5	35.7	95.2
				8	31.3	98.1
				13	18	103.4
B	130.5	0.32	0.13	1	35.7	82.7
				6	27.5	91.4
				13	15.9	103
C	144.0	0.36	0.07	1	28.6	84.8
				6	25.8	95.6
				13	16.3	111.6
D	204.5	0.39	0.05	1	29.6	94
				7	27.1	103
				13	17.6	103.4

Table 5. Mechanical, microstructural and morphological data extracted from squeezed samples in different points.

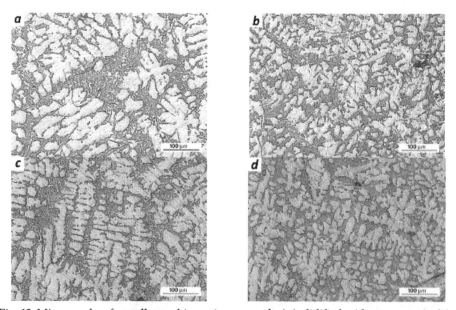

Fig. 13. Micrographs of metallographic sections: sample A (solidified without squeezing) in the middle zone (a), and in the edge (b); sample B (at the squeezing pressure of 100 MPa) in the middle zone (c), and in the edge (d).

Figure 14 shows the experimental curves obtained from microhardness tests (Figure 14-a) and FIMEC tests (Figure 14-b). The higher the pressure, the higher the microhardness: also data scattering reduces due to the squeezing. The pressure has a significant effect on the entire FIMEC curve as it shifts toward higher force values, by increasing pressure.

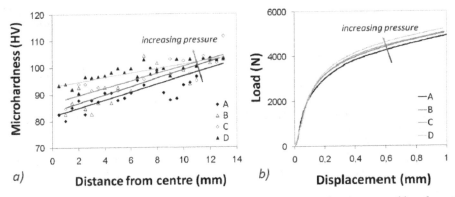

Fig. 14. Mechanical tests for all the squeezing pressures: (a) microhardness profiles along the sample radius; (b) FIMEC tests at the centre.

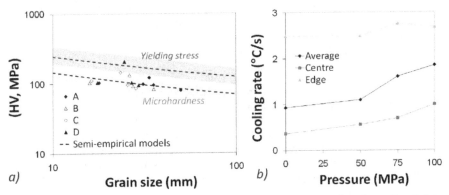

Fig. 15. Theoretical predictions: (a) comparison between experimental and predicted data for mechanical properties; (b) cooling rate inference by microhardness values at different squeezing pressures.

Microhardness is directly correlated to the microstructure. A similar value is measured near the skin for all the squeezing pressures whereas a significant difference is observed toward the centre. Moreover, a higher data scattering is observed at lower pressure values. At 100 MPa, the trend tends to be an horizontal line and a lower dispersion is measured. Microhardness and dendrite size values of Table 5 are evaluated in the same points along the radius. Instead the yielding stress is always evaluated at the centre of the samples. Experimental values for yielding stress and microhardness are reported in Figure 15-a together with the theoretical trends extracted respectively from equation (1) and (2). The semi-empirical models were fitted by using the constants described in the previous section and obtained by squeezing tests on small samples of the same alloy. There is a good agreement between experimental data and predicted ones. A worse agreement is observable for the yielding stress and is probably due to the higher size of the zone interested to the indentation.

As experimental data are in agreement with predictions of equation (1) and (2), it is expected that equation (4) and (5) can be used for the same alloy to infer cooling rate during solidification. In Figure 3-b, microhardness values of Table 5 were used to infer cooling rate at the different squeezing conditions in three sample positions. It is evident that the cooling rate is similar near the edge for all the squeezing pressures.

Cooling rate can be extracted from cast parts by means of a simple indentation test. This is an important operation not only to provide correct inputs for numerical simulations, but above all, to optimize the casting process itself. In fact, it is difficult to define a precise correlation between process parameters and cycle time. Moreover, the cycle time is strongly dependent on the cooling rate. In die casting optimization, performing simple indentation tests on the first cast parts could allow to rapidly converge towards the process optimum. Material equations (4) and (5) were fitted once starting from laboratory specimens but they are applicable on cast products for every casting condition. It is hence possible to characterize all the alloys of interest and use the results to calibrate industrial processes.

5.2.2 Combined effect of the squeezing pressure and the mould temperature

The thermal behaviour of the ingot and the mould in direct squeeze casting is very complex. Sometimes, the experimental evidence could be quite different from the expected trends. In Figure 16, the cast ingots are shown for the nominal mould temperature of 300 °C: the surface appearance is put in evidence together with the typical microstructure. The average roughness for the same ingots is reported in Figure 17. The roughness is expressed in terms of mean value and dispersion for each sample.

In Figure 18, all the results from the mechanical characterization are shown. From the flat indentation tests, the load at the displacement of 0.4 mm was acquired and reported in Figure 18-a in terms of mean value and dispersion for each ingot. Figure 18-b shows the microhardness profile along the cylinder axis for two samples cast at different mould temperature and squeezing pressure. The profile is qualitatively similar with a minimum in the centre. The microhardness mean values and dispersions are reported in Figure 18-c. The hardness increase is strictly dependent on the microstructure refinement (Figure 16).

In the reported experimentation, fixing the mould temperature and the squeezing pressure, the highest ingot cooling rate would be expected for the lower mould temperature as in the initial stage the ingot is cooled by the heat transfer with the steel mould. The mould heating rate should be the lowest as well. But experimental measurements show that the highest mould heating temperature is related to the mould temperature of 350 °C instead of 300 °C. That is due to the transient phase between the beginning of the melt pouring and the pressure application. In those 12 s, the molten metal can suddenly solidify, affecting the heat transfer during the squeezing. As a consequence, best mechanical properties are obtained at the higher mould temperature at which the melt solidification mainly occur during squeezing. This is evident both from FIMEC data (Figure 18-a) and microhardness data (Figure 18-b). At the same mould temperature, mechanical performances generally increase with applied pressure. Data scattering is high due to several occurrences during casting; such as the formation of an oxide coating on the air exposed melt surface or the turbulence of the molten metal during pouring.

Samples	Surface appearence	Micrographs

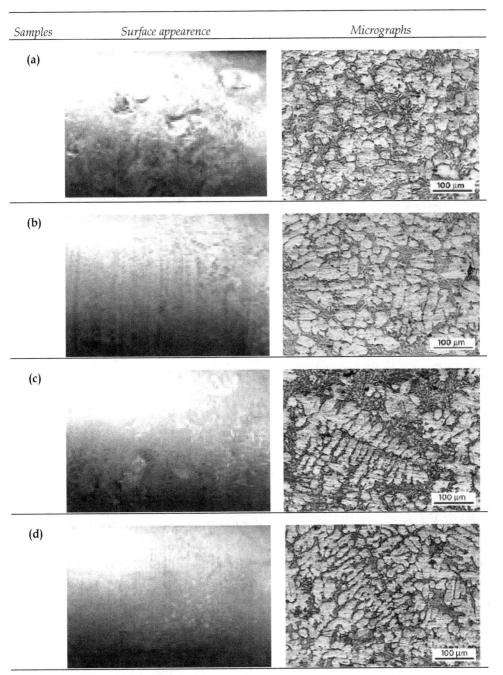

Fig. 16. Surface appearance and micrographs of the ingots produced at the lower mould temperature: samples a, b, c and d.

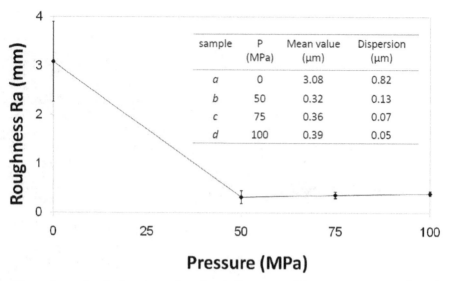

Fig. 17. roughness for the ingots produced at the lower mould temperature: samples a, b, c and d.

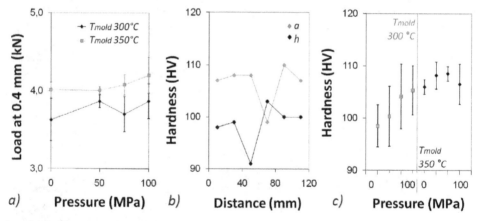

Fig. 18. Experimental results: mean indentation load (a), hardness profile along the cylinder height for the ingots a and h (b), mean hardness for all the ingots (c).

Figure 19-a reports the experimental heating curves of the mould in the first 80 s from the melt pouring: it is evident that higher temperature profiles are reached with lower squeezing pressures; some acquisitions were interrupted before 80 s for the occurrence of technical problems. However, the dependence of the mechanical performances on the final ingot cooling rate is always clear. From the value at 0 s, reported in Figure 19-a, it is possible to measure the real initial mould temperature. In the range between 40 and 80 °C the mould heating rate was extracted and related in Figure 19-b to the mean microhardness of the final ingot.

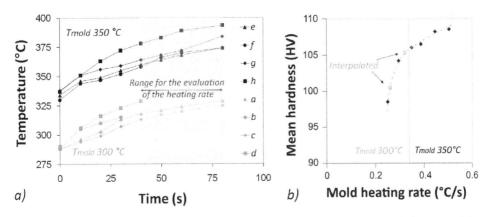

Fig. 19. Experimental results: heating curves of the mould during casting for all the ingots (a), correlation between the mould heating rate and the mean hardness (b).

The increase in the mould heating rate depends on the increase of the ingot cooling rate and therefore is related to the increase in the microhardness. The trend is evident and two points were interpolated to extract the mould heating rate for the castings that gave problems during the temperature acquisition. Higher mould heating rates (i.e. higher ingot cooling rates) lead to higher mechanical properties (Figure 19-b) because of the microstructure refinement (Figure 16). Moreover, the surface properties enhance due to the squeeze as Figures 16 and 17 show in terms of surface appearance and roughness.

5.3 Numerical simulation for the prediction of the ingot mechanical properties

Also in this case, a FE thermal model was defined to investigate the material behaviour during squeezing and to predict mechanical performances of the ingots. The model was calibrated by means of the experimental tests performed in the second set of samples. The FE model geometry was built according to the geometry of Figure 20-a, with a cavity of 28 mm radius and a length of 120 mm. A parametric approach was used by means of a batch procedure to easily modify geometrical and material parameters during the simulations.

In Figure 20-b the FE model is described. A 2D axisymmetric model was used, therefore the hole of the thermocouple was not modelled. The nodes at the interface between the mould and the plug were merged whereas an interface was modelled between the aluminium ingot and the steel mould. The interface was a thin (1 mm) layer which was put from the side of the mould and merged both to the ingot and to the mould. As initial condition, the aluminium ingot temperature was fixed to 750 °C whereas the mould temperature was put equal to the experimental mould temperature (i.e. 290 and 334 °C instead of the nominal values of 300 and 350 °C). A convection load was applied only to the lateral walls and typical values for convection in air were used (the bulk temperature T_b equal to the room temperature and the convection coefficient h equal to 10 W/m² K).

For the steel, general material properties of an AISI steel were used as well as general properties of an aluminium alloy were considered. Only the solidus and liquidus temperature of the aluminium alloy were experimentally measured as they significantly affect the beginning and the end of the solidification. Figure 21 shows the heating and cooling curve of an aluminium alloy sample; this curve was obtained by immerging the thermocouple into the melt. Considering a fusion heat of 390 J/g, a specific heat of 0.88 J/g °C and the acquired solidus and liquidus temperatures, the enthalpy curve of Figure 20-b was constructed.

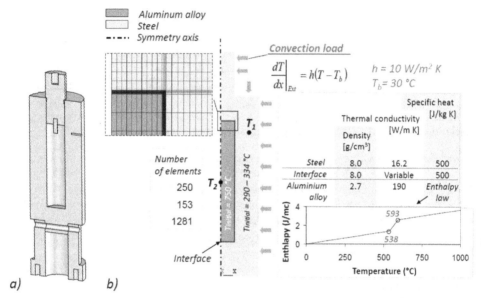

Fig. 20. Numerical modelling: (a) section of the squeezing system; (b) 2D mesh, loads and material properties used in the simulation.

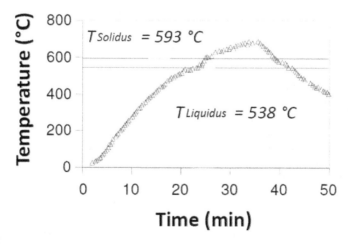

Fig. 21. The heating-cooling curve for the aluminum alloy.

The interface material had the same properties of the steel of the mould except for the thermal conductivity. This datum was used to take into account the different squeezing pressures. As squeezing determines a better matching between the aluminium ingot surface and the mould, this effect can be modelled by increasing the thermal conductivity of the interface material. This assumption allows to simplify the model and to reduce the time for the simulations.

In Figure 20-b there are reported two reference nodes for the temperature acquisition. The first point T_1 represents the thermocouple position in the experimentation. The point T_2 is at the centre of the ingot to evaluate to thermal history of the aluminium alloy during the solidification.

In Figure 22 typical results of the FE model are shown in dependence of the nominal mould temperature and the interface thermal conductivity. The results are expressed in terms of temperature evolution during time for T_1 (Figure 22-a) and T_2 (Figure 22-b). During the ingot cooling, the mould temperature increases up to a maximum due to the heat provided by the ingot solidification (Figure 22-a). Subsequently the temperature decreases due to the convection load. By increasing the nominal mould temperature, the temperature curves are vertically shifted. By increasing the interface thermal conductivity, the temperature peak is shifted at lower times and higher values due to the faster ingot solidification. During the alloy solidification, the ingot temperature decreases following a three stage curve (Figure 22-b). The first stage has a high slope and is representative of the molten metal cooling. At the beginning of the solidification the curve slope decreases. It increases again only at the end of the solidification.

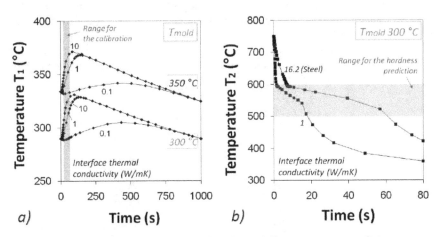

Fig. 22. Temperature evolution for T_1 (a) and T_2 (b) in different casting conditions.

The FE model was used to simulate the ingot cooling under different casting conditions. This goal was obtained by changing the mould temperature and the interface thermal conductivity. For each mould temperature (290 and 334 °C, nominally 300 and 350 °C), the thermal conductivity was changed from 0.1 to 25 W/m K. In order to extract a mould heating rate, the slope of the temperature curve at T_1 (Figure 22-a) was evaluated in the range between 50 and 90 s (as Figure 22-a shows). This numerical datum has to be compared with the experimental datum of Figure 19-a for each casting (taking into account the delay time after pouring, about 10 s). Moreover, from Figure 16, the ingot cooling rate can be extracted for the same combination of mould temperature and interface thermal conductivity. The numerical ingot cooling rate was extracted in the range between 500 and 600 °C (i.e. around the alloy freezing range) as mainly in this temperature range the microstructure of the ingot is formed.

In order to predict the mechanical performances of the final ingots, the real thermal history of each ingot should be taken into account. Considering the experimental mould heating rate, it is possible to extract from Figure 23-a the thermal conductivity data that could provide the same heating rate with the model. Moreover, it is possible to extract from Figure 23-b the mould heating rate which is correspondent to the same thermal conductivity (and so to the same experimental mould heating rate). This way, any experimental occurrence that could affect the thermal behaviour of the ingot is compensated and the effective ingot cooling rate is obtained for each experimental acquisition of the mould heating rate. At the end, the numerical ingot cooling rates can be used in equation (1) to predict the ingot average hardness. In Figure 23-c the comparison between experimental hardness data and predicted data is shown for all the ingots. The figure underlines an optimal agreement between experimental and numerical data. This occurrence shows the suitability of equation (1) for the prediction of mechanical performances of aluminium alloy castings together with the validity of the proposed numerical procedure.

In conclusion it is possible to predict final mechanical properties of direct squeeze cast aluminium alloy parts. In this study, thanks to the simple ingot geometry, a 2D FE parametric model was used. After calibration, it is possible to use the model to infer from the measured mould heating rate, the numerical ingot cooling rate. This correlation has to be made only once and it is not necessary to run the solution for every change of the process parameters. Finally, the numerical ingot cooling rate is used to predict the ingot mechanical properties by using the proposed material equations (1) and (2) or other equivalent material laws.

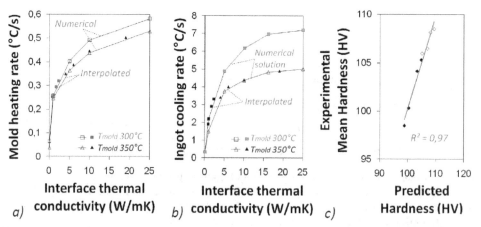

Fig. 23. Numerical results: mould heating rate (a) and ingot cooling rate (b) as a function of the interface thermal conductivity; comparison between experimental and numerical results (c).

Nomenclature

σ_Y	yield strength
σ_U	ultimate tensile strength
HV	Vickers hardness
σ_0, K_Y	constants of Hall-Petch equation for yield strength
HV_0, K_H	constants of Hall-Petch equation for hardness
λ	average dendrite cell size
ε	cooling rate
B, n	alloy specific constants in the relationship between λ and ε
C_Y, m	constants in the relationship between σ_Y and ε
C_H, m	constants in the relationship between HV and ε
q_Y, q_S	limit and saturation load values during FIMEC test
T	temperature
h, T_b	heat transfer coefficient and bulk temperature for convective load
R^2	correlation coefficient for linear interpolation

6. References

Britnell, D.J., Neailey, K. (2003) *Journal of Materials Processing Technology* 138 306-310.

Donato, A., Gondi, P., Montanari, R., Moreschi, L., Sili, A. and Storai S. (1998) "A remotely operated FIMEC apparatus for the mechanical characterization of neutron irradiated materials" *J.Nucl. Mater.* Vol. 258-263 pp. 446-451.

Gallerneault, M., Durrant, G. and Cantor, B. (1995) "Eutectic channelling in a squeeze cast Al-4.5wt%Cu alloy", *Scripta Metallurgica et Materialia*, Vol.32, No.10, pp.1553-1557.

Gallerneault, M., Durrant, G. and Cantor, B. (1996) "The squeeze casting of hypoeutectic binary Al-Cu", *Metallurgical and Materials Transactions A*, Vol.27A, pp.4121-4132.

Gondi, P., Montanari, R., Sili, A., (1996) "Small scale penetration tests with cylindrical indenters" *Proceedings of IEA international symposium Julich Germany*, pp.79-84.

Han, Y.S., Kim, D.H., Lee, H.I. and Kim, Y.G. (1994) "Effect of applied pressure during solidification on the microstructural refinement in an Al-Cu alloy", *Scripta Metallurgica et Materialia*, Vol.31, No.12, pp.1623-1628.

Hong, C.P., Shen, H.F. and Lee, S.M. (1998) "Prevention of macrodefects in squeeze casting of an Al-7 wt pct Si alloy", *Metallurgical and Materials Transactions B*, Vol.31B, pp.297-305.

Hong, C.P., Shen, H.F. and Cho, I.S. (1998) "Prevention of macrosegregation in squeeze casting of an Al-4.5 wt pct Cu alloy", *Metallurgical and Materials Transactions A*, Vol.29A, pp.339-349.

Kalpakjian, S. (2000), *Manufacturing processes for engineering materials – 3° Edition*, Addison-Wesley Publishing Company, ISBN 0-201-30411-2.

Kim, S.W., Durrant, G., Lee, J.H. and Cantor, B. (1999) "The effect of die geometry on the microstructure of in direct squeeze cast and gravity die cast 7050 (Al-6.2Zn-2.3Cu-2.3Mg) wrought Al alloy", *J. of Materials Science*, Vol.34, pp.1873-1883.

Lee, J.H., Won, C.W., Cho, S.S., Chun, B.S. and Kim, S.W. (2000) "Effects of melt flow and temperature on the macro and microstructure of scroll compressor in direct squeeze casting", *Materials Science and Engineering*, Vol.A281, pp.8-16.

Lee, J.H., Kim, H.S., Won, C.W., Cantor, B. (2002) *Materials Science & Engineering* A 338 182-190.

Maeng, D.Y., Lee, J.H., Won, C.W., Cho, S.S. and Chun, B.S. (2000) "The effects of processing parameters on the microstructure and mechanical properties of modified B390 alloy in direct squeeze casting", *J. of Materials Processing Technology*, Vol.105, pp.196-203.

Reed Hill, R.E. (1996) *Physical metallurgy principles*, third edition. PWS publishing Company, Boston. pp.192.

Savas, M.A., Erturan, H. and Altintas, S. (1997) "Effects of squeeze casting on the properties of Zn-Bi monotectic alloy", *Metallurgical and Materials Transactions A*, 1997, pp.1509-1515.

Shuangjie, C. and Renjie, W. (1999) "The structure and bending poperties of squeeze-cast composites of A356 aluminium alloy reinforced with alumina particles", *Composites Science and Technology*, Vol.59, pp.157-162.

8

Die Casting and New Rheocasting

Matjaž Torkar[1], Primož Mrvar[2], Jožef Medved[2],
Mitja Petrič[2], Boštjan Taljat[3] and Matjaž Godec[1]
[1]Institute of Metals and Technology, Ljubljana,
[2]University of Ljubljana, Faculty of natural sciences and engineering,
Department of materials and metallurgy, Foundry chair, Ljubljana,
[3]Steel d.o.o., Ljubljana,
Slovenia

1. Introduction

1.1 The high-pressure die casting process

The high-pressure die casting process (HPDC) is a rapid solidification process leading to formation of rapid solidified castings. There are some specifics of the process. The casting of a molten alloy into a mold is complete within several milliseconds. A significant quenching effect and a high production rates are possible. The application of high-pressure enables good contact between molten alloy and die wall that enables: the increase in cooling rate, the increase in heat flow and heat transfer coefficient at the die-melt interface as well as the formation of a net shape casting. Casting defects such as shrink holes which generate by the shrinkage during solidification are reduced. Components with complicated shapes are produced directly from a liquid state even for a molten alloy with high viscosity.

By taking these facts into account, it is expected that much larger shapes and dimensions can be formed in various alloy systems by the high pressure die casting process.

The characteristics of HPDC process (Andersen, 2005, Vinarcik 2003), are high velocity of melt during filling the die and high solidification rate of the component. Such circumstances demands more sophisticated approach to the study of phenomena during HPDC process.

Among others an integrated virtual and rapid prototyping methodology is proposed for advanced die-casting manufacturing using a hot-chamber process (Ferreira et al., 2007). This approach enabled optimization of the die-casting manufacturing technology parameters and reduced the lead-time of die-casting designs.

The physical, mechanical and esthetical properties of the components are directly dependent on process conditions during casting; the die temperature, the metal velocity at the gate, the applied casting pressure, the cooling rate during die casting, the geometrical complexity of the component and the mold filling capacity. All that affects also the integrity of the cast components. If these parameters are not controled properly, various defects within the finished component may be expected. The applied casting pressure is crucial during the solidification of high integrity parts. The effects of process variables on the quality of the cast components with in-cavity pressure sensors delay time and casting velocity were examined by Dargusch in 2006. He found that the porosity decreases with increasing pressure and increases with higher casting velocity (Dargusch et al., 2006, Laws et al., 2006).

The latest development of rheocasting process is based on the principles of HPDC. The material in semi-solid state is pressed into the tool cavity. That enables faster solidification and better productivity. Due the narrow temperature window the optimization of the new rheocasting process demands precise control of process parameters.

Presented is development of the microstructure of the semi-solid slurry and the rheocast component and an overview of observed defects in components produced by new rheocating process. A comparison between the simulated porosity and experimental determined porosity is presented.

1.2 Evolution of the microstructure at new rheocasting process

Evaluation of the microstructure of the components, manufactured by rheocasting proces, revealed the outgrowths on the surface of primary globular α_{Al} crystall grains. The outgrowths can grow into dendrites that are not suitable for the new rheocasting process. The explanation of these phenomenon is in the theory of solidification.

The alloys with dendritic microstructure in two phase region are not suitable for the new rheocast process, because dendritic solidified material has no isotropic properties. Discovery of Flemings, that material with globulitic microstructure in two phase region (L+α) behaves thixotropic, enabled the further development of hot working in semisolid state (Giordano at all 2002, Müller-Späth at all, 1997, Wabusseg at all, 2002, Kaufmann at all, 2001, Hall at all, 2000, Curle at all, 2010, Curle at all, 2011, Cabibbo at all, 2001, Kapranos at all, 2001) like thixocasting and new rheocasting (NRC) (Sereni, 2005). The efforts to introduce the hot working in semi solid state (Blazek at all, 1995) exists also for steels but these efforts are less sucessful due higher temperatures.

Hot working in semi solid state bases on preparation of wrouth material with more or less globulitic forms of solidified primary phase, surrounded by molten material. Free Gibb's molar energy of globular form is presented by equation (1).

$$\Delta G_L = \frac{2\gamma}{R} \tag{1}$$

Where R represents radius of bent interphase boundary and γ interphase free energy between liquid/solid. Evident is the equality with the classical theory of nucleation, where the free energy, necessary for formation of nucleus with critical radius R*, equals (2)

$$\Delta G_\gamma = \frac{2\gamma}{R^*} \tag{2}$$

In the case of equilibrium among the phases solid/liquid and due the influence of curvature of solid phase after Gibbs-Thomson the ΔG_L equals ΔG_γ. Taking into account this balance and with supposition that specific surface energy of solid phase is isotropic, than from both expression for ΔG_L and ΔG_γ follows that free energy is proportional to the lowering of the temperature, as shown in equation (3)

$$\Delta T_R = \frac{2\gamma}{\Delta S}\frac{1}{R} \tag{3}$$

Where ΔS is the difference in entrophy solid/liquid, R is radius of curvature of interphase surface and γ is free energy of the surface solid/liquid. Round nucleus is stable until $R > R^*$ and are remelted when $R<R^*$.

At the begining of the growth the nucleus is stable $R >R^*$ and possible are only smal outgrowths with lower radius $r <<R^*$. The outgrowths remelt in the case that $\Delta T_R > | \Delta T_L + \Delta T_C|$, where ΔT_R express the undercooling due the curvature of the outgrowth, ΔT_L is undercooling due the local temperature of the melt, ΔT_C is the contribution of constitutional undercooling because of the melt composition. Until the ΔT_R is high enough the nondendritic globulites can grow. With increasing size of solid globules, the larger outgrowths on the surface are possible and they can develop further into dendrites.

As the primary phase α_{Al} grow further in semi solid region (mixture of solid-liquid state) the importance of the $| \Delta T_L + \Delta T_C|$ increases. For the further growth of globulites both, ΔT_L and ΔT_C must be reduced, which is possible either by forced mixing of the melt or with slower cooling rate.

The NRC process bases on lower cooling rate during the growth of primary phase in the preform (slurry). Due the influence of mixing and diffusion at slower cooling rates, the distribution of solid phase, near the contact solid/liquid is more equal compared to distribution of solid phase at rapid cooling of the melt. This enables lower constitutional undercooling ΔT_C. At lower ΔT_C the Gibbs-Thompsons effect increases the stability of the interphase surface. Thus, lower cooling rate accelerates the growth of globulites (Zhu at all, 2001, Uggowitzer at all, 2004).

The cells and dendritic solidification is a consequence of constitutional undercooling which means the temperature of the melt at solidifying front is lower compared to equilibrium solidification.

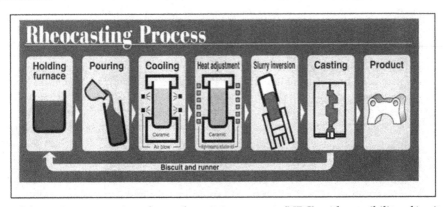

Fig. 1. Schematic presentation of new rheocasting process (NRC) with possibility of in situ recycling of material (From project documentation GRD1-2002-40422).

The formation of globulitic structure in the slurry demands first high cooling rate to get a high number of small nucleuses and after that to cool down the melt slowly (Sereni, 2005).

In comparison with thixocasting (Kapranos at all, 2000) the new rheocasting process (NRC), shematic presented in Fig. 1, has some additional benefits: continuous casting with electromagnetic stirring is not necessary, there is no cutting of the billets and no reheating of the cut pieces into the semi solid region for the forming in the die. NRC process (Sereni, 2005, Torkar at all, 2005) starts with liquid phase, followed by preparation of semisolid slurry that is directly transfered into the die of the press.

The main aim of the investigation was evaluation of the microstructure and presentation of the most often failures in real automotive component, manufactured by NRC process.

2. New rheocasting process

2.1 Material and preparation of slurry

Metallographic investigations of the microstructure of slurry (partly solidified preform, before the entrance into the press) and the component, both of hypoeutectic silumin A 357 were performed. Typical compositon of the alloy is presented in Table 1. The components were high pressure die cast on NRC device and then heat treated (T5, 6 hours at 170 $^{\circ}$C).

Cu	Mg	Si	Fe	Mn	Ti	Zn	Sr	Al
0.2	0.4 – 0.7	6.5 – 7.5	Max. 0.2	Max. 0.2	0.05 –0.2	Max. 0.2	0.03	Rest

Table 1. Tolerances of chemical composition of A 357 alloy in wt. %.

The samples for metallographic investigations were cut from the slurry and from the component and prepared by standard metallographic procedure. The soundness of the components was checked by x-ray device. The detected failures were checked by metallography.

The microstructure was observed by the light microscope Nikon Microphot FXA, equiped with 3CCD- videokamero Hitachi HV-C20A in software analySIS for evaluation of the microstructure. The hardnes of the component was measured by Brinell.

2.2 Microstructure characterisation and defects presentation

The microstructure of the middle of the slurry is presented in Fig. 2a, b. We can observe the globules of primary phase α_{Al} and equally distributed eutectic among the globules. No other specifics were observed in the microstructure of the slurry. Similar microstructure was observed also at high pressure die cast component. The globules of primary phase α_{Al} are enveloped with eutectic (Fig. 3). Beside that, especially in thinner regions of the component, we can observe fingers like outgrowths (start of grow of dendrites) on the surface of globulites (Fig. 4).

There are two possibilities for the appearance of these outgrowths. First possibility is local change in undercooling, during the deformation in the die that enables the appearance of dendrites. The second possibility is not yet finished globularisation during cooling of slurry. More likely is the explanation with local change of undercooling and its influence on distribution of solidified phase during the process of deformation in the die. It is well known that higher undercooling and distribution of solid phase accelerates the formation of dendrites. The second possibility is denied based on microstructure (Fig. 2b), where no outgrowths were observed and beside that the slurry was cooled down in more controlled temperature regime.

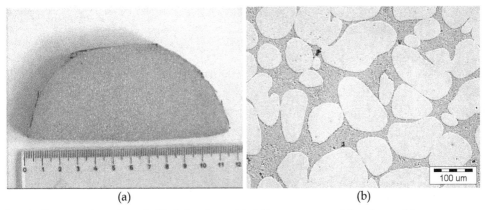

(a)   (b)

Fig. 2. Cross section of one half of the slurry (a). Microstructure of the slurry (b).

The explanation of the appearance of the outgrowths and dendrites, observed in the microstructure of the component one can get in the theory of solidification. Until the contribution of ΔT_R is high enough, the surface of the growing nucleus is stable and the growth of globule proceeds. When the size of solid globule increases, increases also the possibility of development of outgrowth with higher radius that can degenerates into dendrite.

Studies (Zhu at all, 2001, Uggowitzer at all, 2004, Dahle at all, 2001) of the concentration profiles of solute in the melt during formation of the primary phase at different cooling rates, revealed that at higher cooling rates increases the concentration of solute in the melt near the S/L interface due redistribution of the melt on the interface and due short time for diffusion of the solute. Stability of the interface is fast destroyed. That enables the appearance of outgrowths and start of the dendrite grows on the surface of the primary phase α_A, (Fig. 5), as observed in the investigated component.

Lower casting temperature of slurry combined with the lower cooling rate, accelerates the formation of globular micro structure. The formation of the globular structure is accelerated also by the inoculation of the melt.

In the microstructure of the component beside globules also some dendrites of the primary phase were observed. The presence of dendrites means that temperature regime of NRC process was not optimal. After the solidification the component was cooled in water and heat treated (T6).

The hardness of the component after heat treatment T6 was among 92-96 HB 5/250.

x-ray examination of the component revealed internal errors like shrinkage porosity (Fig. 6a, b), gas cavities (Fig. 7a, b) and large inclusion (Fig. 8a, b) that were confirmed also by metallography. Observed were also other irregularities like not filled form, overcasts on the surface, segregations, precipitates of silicon and dendrite formation (Fig. 9a-f).

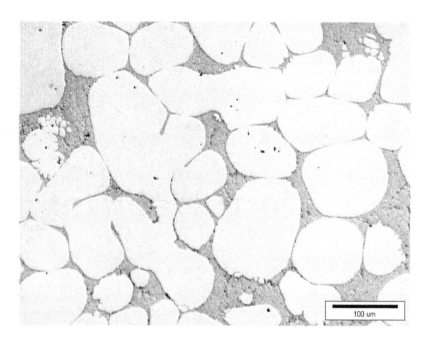

Fig. 3. Microstructure of the component.

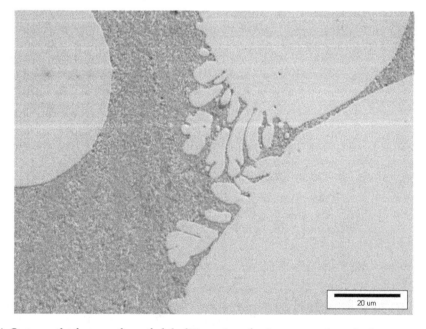

Fig. 4. Outgrowths from surface of globulitic grains of primary α_{Al} phase in the component.

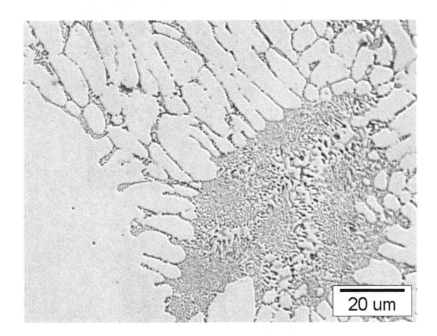

Fig. 5. Dendrites of primary α_{Al} phase in the component.

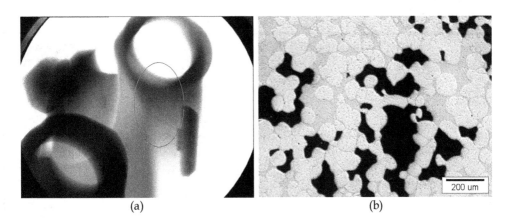

(a)   (b)

Fig. 6. (a) Radiograph of internal defect in the component (b) Central shrinkage porosity confirmed by metallography.

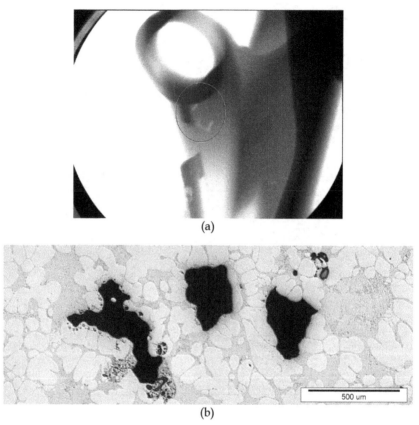

(a)

(b)

Fig. 7. (a) Radiograph of internal defects (b) Combination of shrinkage and gas porosity, confirmed by metallography.

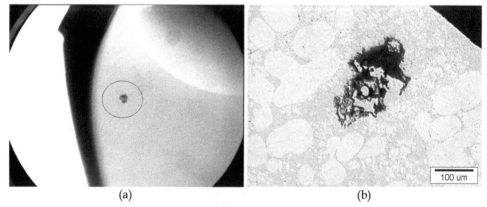

(a)   (b)

Fig. 8. (a) Radiograph of internal defect (b) Oxide inclusion, confirmed by metallography.

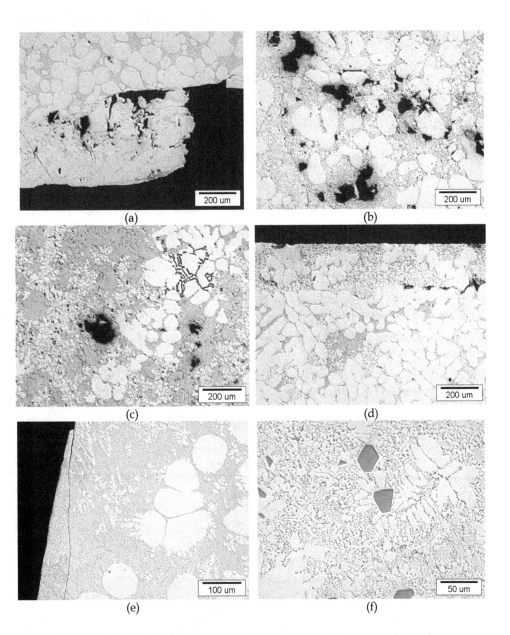

Fig. 9. (a) Not filled surface in the component. (b) Central shrinkage porosity in the component. (c) Eutectic, segregation and porosity. (d) Overcast near the surface of the component. (e) Cold weld on the surface of the component. (f) Microstructure of investigated alloy contains primary crystals α_{Al} dendrites, eutectic and non equilibrium primary crystals of β Si.

Based on metallographic examinations of the rheocast components we can observe the presence of several typical surface and internal defects that should be eliminated. Necessary is more precise control of parameters of preparation of slurry from the melt and at pressing and solidification in the die.

2.3 Conclusions on rheocasting

The microstructure of the slurry is homogenous and with equal distribution of primary solidified globular grains.

Compared to slurry, more irregularities in the microstructure were observed in the high pressure die cast components.

The formation of outgrowths on globular grains of primary phase α_{Al} and formation of dendrites is a consequence of local changes in undercooling in the semisolid state, during deformation of material in the die.

In the rheocast component the following failures were observed: the primary α_{Al} phase in the form of dendrites, internal defects as central macro- and micro porosity, gas porosity, segregations and inclusions.

On the surface the typical defects were not filled edges of thin wall regions, overcasts, cracks and appearance of blisters during heat treatment.

All these defects show that the temperature regime of investigated new rheocasting process was not yet optimal.

Further R&D is necessary to reduce internal defects by optimization of new rheocasting process parameters, to increase yield and to reduce production costs per component.

3. HPDC shrinkage simulations

The process of high pressure die casting (HPDC) was developed for manufacturing of a large variety of products with high dimensional accuracy. The process is faster and enables more economical production of aluminium automotive components (Dargusch et al., 2006). The rapid development of numerical simulation technology and the solidification simulation of casting has been taken as an effective tool for modeling the casting process and improving the quality of casting (Vijayaram et al., 2006, Dobrzansky et al., 2005). The use of simulation software saves time and reduces costs for the casting system design and the use of materials.

The porosity of the castings can be studied with destructive testing as visual check after machining and non-destructive testing as x-ray microscopy and image processing technology which can provide more detailed information of the gas pores and shrinkages. It is also observed that the chemical composition of the alloy affects the porosity in the cast components, grain refinement and modification (Cleary at al., 2006, Petrič et al., 2011). Now it is commonly accepted that the shrinkage and the gas entrapment are two major causes of porosity. The shrinkage porosity is associated with the "hot spots" in the casting. The gas porosity is caused by entraped air in the injection system and cavity, gas generated from burned lubricants, water in the cavity and hydrogen gas. The entrapped air is unwanted product of high velocity of alloy caused by turbolent flow during injection process.

Presented are results of simulation of the HPDC of Al-Si9Cu3 casting in the H13 steel die and comparison between the simulated and experimental porosity.

3.1 Material and the casting system

The alloy used for die casting was aluminium-silicon-copper alloy (Table 1), marked by less affinity to shrinkage and internal shrinkage cavities and very good castability.

Si	Cu	Fe	Mn	Mg	Zn	Ni	Cr
10.38	2.73	0.82	0.25	0.34	0.82	0.04	0.04

Table 2. Chemical composition of Al-Si9Cu3 alloy in wt %.

ALSI H13 chromium hot work tool steel was used for the die. This steel has higher resistance to heat cracking and die wear caused by the thermal shock associated with the die casting process. The casting system with a shot sleeve and a plunger are presented on Figure 10. Presented are gates and runner system with two cavities and the final product, an automotive component.

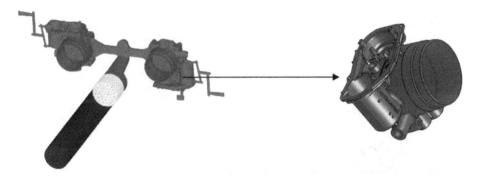

Fig. 10. Casting system; shot sleeve with plunger, gates and runner system, two cavities and the casting component.

3.2 Characteristics of HPDC process

The casting process is divided into four phases: pre-filling, shot, final pressure phase and ejection phase. In the pre-filling phase, the molten metal is injected by plunger, which forces the metal with a low velocity through a horizontally mounted cylindrical shot sleeve up to the gate. The shot sleeve is usually partially filled with molten metal, in an ammount that depends on the volume of the cast component. The fluid flow and the amount of empty space are affected by plunger motion, shot sleeve dimensions and amount of metal in the sleeve (Thorpe et al., 1999). In short shot phase the plunger is accelerated to high velocity and sufficient venting of the die cavity is practically impossible. In the final pressure phase, solidification of the casting is completed and in the ejection phase, moulded part is removed, die halves are sprayed and positioned back to repeat the cycle.

The industrial HPDC process started with plunger, that has four different speeds, as it is shown on shot profile in Fig. 11a. The volume fraction in Fig. 11b shows that no wave and no air entrapement were formed.

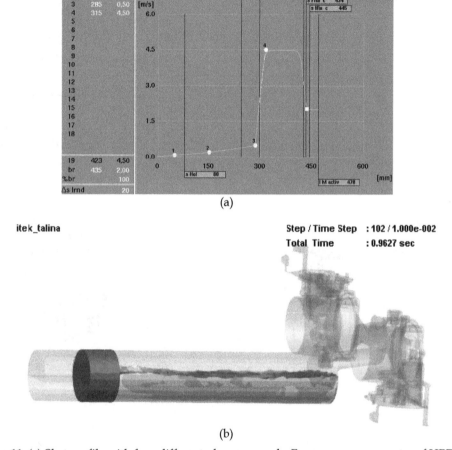

(a)

(b)

Fig. 11. (a) Shot profile with four different plunger speeds. From process computer of HPDC machine. (b) Volume fraction picture of alloy and empty space in the shot sleeve.

3.3 The shot sleeve simulation

The movement of the plunger was simulated by three different plunger speeds with FEM-based software ProCast. The simulation is shown on shot profile in Fig. 12a, b. The volume fraction in Fig. 12c shows no wave and no air entrapement.

The set up time was minimised, the plunger speed optimized and the industrial HPDC process was shortened for 0,48 s.

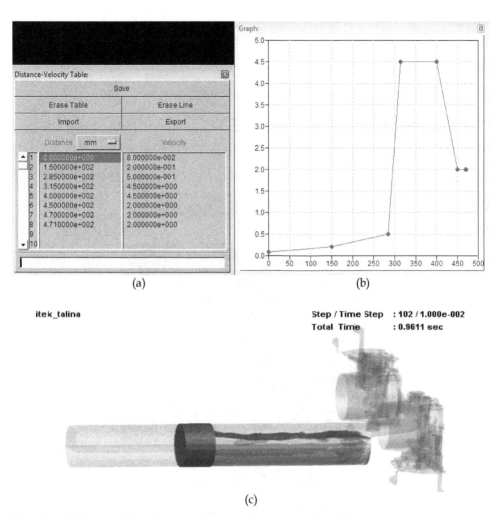

(a)   (b)

(c)

Fig. 12. (a,b) Shot profile with three different plunger speeds. (c) Volume fraction picture of alloy and empty space in the shot sleeve.

3.4 The shrinkage porosity simulations

The shot sleeve simulation results were used as boundary condition for cavity filing simulations and shrinkage porosity simulations. Basic study in this paper was the shrinkage porosity. Fig. 13a, b shows simulated shrinkage porosity "red spots" in left and right casting. After 9 cycles of casting the constant conditions on the die were established and after 10 cycle we cut the left side casting to examine two red spot of simulated shrinkage porosity (Fig. 14a,b and 15a,b).

The two cross sections of left casting (Fig. 14b, 15b) show good agreement with the simulated results of shrinkage porosity (Fig. 13a, 14a and 15a).

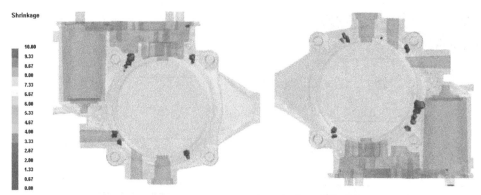

Fig. 13. (a, b) Shrinkage porosity simulation on left and right castings.

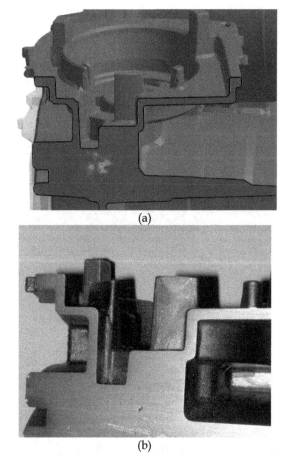

Fig. 14. (a, b) Shrinkage porosity in left casting: simulation and cross section.

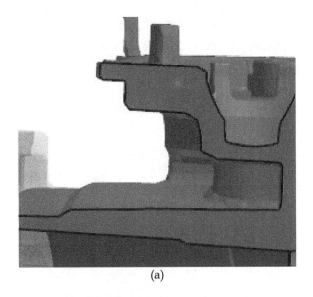

(a)

(b)

Fig. 15. (a, b) Shrinkage porosity in left casting: simulation and cross section.

3.5 Conclusions on porosity simulations

In the present work the porosity of automotive components has been analyzed with ProCast, FEM-based software. The most important conclusions obtained are:

-   The shot sleeve simulation gives valuable information for the final quality of the components by minimising the volume fraction of the empty space during the first stage of the HPDC process. The volume fraction shows no wave and no air entrapement.
-   The shot sleeve simulation gives savings in lead time by minimising the set up time during the shot stage of HPDC process. The shot stage of HPDC process set up time was shortened for 0,48 s.
-   The shot sleeve simulation gives also information of the location of the shrinkage porosity in castings, called "red spots". The shrinkage porosity in cross section of real automotive component show good agreement with the results of simulations.

4. Acknowledgment

A part of this work was funded primarly by G1RD-CT-2002-03012 grant of EU.

5. References

Andersen B., *Die casting Engineering: a hydraulic, thermal and mechanical process*, Marcel Dekker, 2005

Ferreira J.C., Bartolo P.J.S., Fernandes Alves N.M., Marques J., Virtual and rapid prototyping for rapid die-casting development, *International Journal of Computer Applications in Technology*, vol. 30, no. 3, 2007, pp 176-183

Blazek K.E., Kelly J.E., Pottore N.S., The Development of a Continuous Rheocaster for Ferrous and High Melting Point Alloys, *ISIJ International,* Vol.35, (1995), 6, 813-818

Cabibbo M., Evangelista E., Spigarelli S., Cerri E., Characterisation of a 6082 aluminium alloy after thixoforming, *Materiali in tehnologije* 35 (2001) 1-2, 9-16

Cleary P.W., Ha J., Prakash M., Nguyen T., 3D SPH flow predicitons and validation for high pressure die casting of automotive components, *Appl. Math. Model.*, 30 (2006) 1406-1427

Curle U.A., Semi-solid near-net shape rheocasting of heat treatable wrought aluminium alloys, *Transactions of Nonferrous Metals Society of China* (English Edition), vol. 20, no. 9, pp. 1719–1724, 2010

Curle U.A.,, Wilkins J.D., Govender G., Industrial Semi-Solid Rheocasting of Aluminum A356 Brake Callipers, *Advances in Materials Science and Engineering*, Volume 2011 (2011), Article ID 195406, 5 pages, doi:10.1155/2011/195406

Dahle A.K, Nogita K., Zindel J.W., McDonald S.D., Hogan L.M., Eutectic Nucleation and growth in Hypoeutectic Al-Si Alloys at Different Strontium Levels, *Metallurgical and Materials Transaction A*, vol. 32A, April (2001), 949

Dargusch M.,S., Dour G., Schauer N., Dinnis, G. Savage G., The influence of pressure during solidification of high pressure die cast aluminium during solidification of high components, *J. Mater. Process. Technol.,* 180 (2006) 37-43

Dobrzanski L.A., Krupinski M., Sokolowski J.H., Computer aided classification of flaws occurred during casting of aluminium, *J. Mater. Process. Technol.,* 167 (2005) 456-462

Giordano P., Chiarmetta G.L., Thixo and Rheo Casting: Comparison on a High Production Volume Component, Proceedings of the 7th S2P Advanced Semi-solid Processing of Alloys and Composites, ed. Tsutsui, Kiuchi, Ichikawa, Tsukuba, Japan, (2002), 665-670

Hall K., Kaufmann H., Mundl A., Detailed Processing and Cost Consideration for New Rheocasting of Light Metal Alloys, Proceedings 6 th International Conference on Semi-Solid Processing of Alloys and Composites, ed. G. L.Chiarmetta, M. Rosso, Torino, (2000), 23-28

Kapranos P., Ward P.J., Atkinson H.V, Kirkwood D.H., Thixoforming – a near net shaping process, *Materiali in tehnologije,* (2001) 1-2, 27-30

Kaufmann H., Uggowitzer P. J., The Fundamentals of the New Rheocasting – Process for Magnesium Alloys, *Advanced Engineering Materials,* 3 (2001) 12, 963-967

Laws K.J., Gun B., Ferry M., Effect of the die-casting parameters on the production of high quality bulk metallic glass samples, *Mater. Sci. Eng.,* A425 (2006) 114-120

Müller-Späth H., Sahm P.R., Razvojni dosežki na področju „Thixocasting" postopka na livarskem inštitutu RWTH Aachen, *Livarski vestnik,* 44, (1997), 2, 33-42

Petrič M., Medved J., Mrvar P., Effect of grain refinement and modification of eutectic phase on shrinkage of AlSi9Cu3 alloy, *Metalurgija,* 50 (2011) 2, 127-13

Sereni S., Rheo-light Final Report, Centro Richerche Fiat, Doc. 2005-133-RP-021, 11.05.2005

Thorpe W., Ahuja V., Jahedi M., Cleary P., Stokes N.: Simulation of fluid flow within the die cavity in high pressure die casting using smooth particle hydrodynamics, Trans. 20th int. die casting cong. & expo NADCA, Cleveland, 1999, T99-014

Torkar M., Breskvar B., Godec M., Giordano P., Chiarmetta G., Microstructure evaluation of an NRC processed automotive component, *Materiali in tehnologije* 39 (2005) 6, 73-78

Uggowitzer P.J., Kaufmann H., Evolution of Globular Microstructure in New Rheocasting and Super Rheocasting Semi-Solid Slurries, *Steel research int.* 75 (2004) No. 8/9, 525-530

Vijayaram T.R., Sulaiman S., Hamuda A.M.S., Numerical simulation of casting solidification in permanent metallic molds, *J. Mater. Process. Technol.,* 178 (2006) 29-33

Vinarcik J.E., *High integrity die casting processes,* John Wiley & Sons, 2003, ISBN: 0-471-20131-6, New York

Wabusseg H., Kaufmann H., Wahlen A., Uggowitzer P. J., Theoretische Grundlagen und praktische Umsetzung von New Rheocasting von Al-Legierungen, *Druckguss-Praxis,* 1 (2002),16-19

Zhu M.F., Kim   J.M., Hong C.P., Modeling of Globular and Dendritic Structure
    Evolution in Solidification of an Al–7mass %Si Alloy, *ISIJ Intern.*, 41 (2001), No.9,
    992-998

Permissions

The contributors of this book come from diverse backgrounds, making this book a truly international effort. This book will bring forth new frontiers with its revolutionizing research information and detailed analysis of the nascent developments around the world.

We would like to thank Mohammad Nusheh, PhD in Extractive Metallurgy, for lending his expertise to make the book truly unique. He has played a crucial role in the development of this book. Without his invaluable contribution this book wouldn't have been possible. He has made vital efforts to compile up to date information on the varied aspects of this subject to make this book a valuable addition to the collection of many professionals and students.

This book was conceptualized with the vision of imparting up-to-date information and advanced data in this field. To ensure the same, a matchless editorial board was set up. Every individual on the board went through rigorous rounds of assessment to prove their worth. After which they invested a large part of their time researching and compiling the most relevant data for our readers. Conferences and sessions were held from time to time between the editorial board and the contributing authors to present the data in the most comprehensible form. The editorial team has worked tirelessly to provide valuable and valid information to help people across the globe.

Every chapter published in this book has been scrutinized by our experts. Their significance has been extensively debated. The topics covered herein carry significant findings which will fuel the growth of the discipline. They may even be implemented as practical applications or may be referred to as a beginning point for another development. Chapters in this book were first published by InTech; hereby published with permission under the Creative Commons Attribution License or equivalent.

The editorial board has been involved in producing this book since its inception. They have spent rigorous hours researching and exploring the diverse topics which have resulted in the successful publishing of this book. They have passed on their knowledge of decades through this book. To expedite this challenging task, the publisher supported the team at every step. A small team of assistant editors was also appointed to further simplify the editing procedure and attain best results for the readers.

Our editorial team has been hand-picked from every corner of the world. Their multi-ethnicity adds dynamic inputs to the discussions which result in innovative outcomes. These outcomes are then further discussed with the researchers and contributors who give their valuable feedback and opinion regarding the same. The feedback is then collaborated with the researches and they are edited in a comprehensive manner to aid the understanding of the subject.

Apart from the editorial board, the designing team has also invested a significant amount of their time in understanding the subject and creating the most relevant covers. They scrutinized every image to scout for the most suitable representation of the subject and create an appropriate cover for the book.

The publishing team has been involved in this book since its early stages. They were actively engaged in every process, be it collecting the data, connecting with the contributors or procuring relevant information. The team has been an ardent support to the editorial, designing and production team. Their endless efforts to recruit the best for this project, has resulted in the accomplishment of this book. They are a veteran in the field of academics and their pool of knowledge is as vast as their experience in printing. Their expertise and guidance has proved useful at every step. Their uncompromising quality standards have made this book an exceptional effort. Their encouragement from time to time has been an inspiration for everyone.

The publisher and the editorial board hope that this book will prove to be a valuable piece of knowledge for researchers, students, practitioners and scholars across the globe.

List of Contributors

Marián Schwarz
Technical University in Zvolen, Faculty of Ecology and Environmental Sciences, Slovak Republic
Institute of Chemistry, Center of Excellence for White-green Biotechnology, Slovak Academy of Sciences, Nitra, Slovak Republic

Vladimír Lalík
Technical University in Zvolen, Faculty of Ecology and Environmental Sciences, Slovak Republic

Mirela Ioana Iorga, Raluca Pop, Marius Constantin Mirica and Doru Buzatu
National Institute for Research & Development in Electrochemistry and Condensed Matter – INCEMC – Timisoara, Romania

F. R. Carrillo-Pedroza, M. J. Soria-Aguilar, E. Salinas-Rodríguez, A. Martínez-Luevanos, T. E. Pecina-Treviño and A. Dávalos-Sánchez
Autonomous University of Coahuila, Mexico

Eetu-Pekka Heikkinen and Timo Fabritius
University of Oulu, Finland

Mihai Ovidiu Cojocaru, Niculae Popescu and Leontin Drugă
"POLITEHNICA" University, Bucharest, Romania

H. Vladimir Martínez
Institute of Energy, Materials and Environment, School of Mechanical Engineering, Pontificia Bolivariana University, Medellín, Colombia

Marco F. Valencia
Engineering School of Antioquia, Medellín, Colombia

Bellisario Denise, Costanza Girolamo, Tata Maria Elisa, Quadrini Fabrizio and Santo Loredana
Department of Mechanical Engineering, University of Rome Tor Vergata, Italy

Boschetto Alberto
Department of Mechanics and Aeronautics, University of Rome La Sapienza, Rome, Italy

Boštjan Taljat
Steel d.o.o., Ljubljana, Slovenia

Matjaž Torkar and Matjaž Godec
Institute of Metals and Technology, Ljubljana, Slovenia

Primož Mrvar, Jožef Medved and Mitja Petrič
University of Ljubljana, Faculty of Natural Sciences and Engineering, Department of Materials and Metallurgy, Foundry Chair, Ljubljana, Slovenia

Printed in the USA
CPSIA information can be obtained
at www.ICGtesting.com
JSHW011357221024
72173JS00003B/322